Praise for *How Your Teen* 
Brain

"While the teenage brain is uniquely sensitive and therefore vulnerable to environmental factors, it is also malleable and adaptive. Indeed, adolescence can be a window of opportunity to bolster brain power. By providing parents concrete steps to support, protect, and integrate the brain, Dr. Coates' new book gives parents a powerful gift. Use it wisely. "

— Victoria Dunckley, MD, child psychiatrist and author of *Reset Your Child's Brain*

"Dr. Coates takes the complicated subject of neuroscience and explains in very practical terms how brain development works as well as how development can be compromised with troubling societal trends–drugs, alcohol and excessive screen time. As a business leader I have seen first-hand both results in the workplace–very developed and poorly developed brains. The dissimilarities make all the difference in what the future holds for our youth. As our school systems seem to be focused on "teaching to the test" rather than critical and creative thinking skills, this practical how-to guide provides parents, grandparents and teens a roadmap for developing the most important asset, the brain, for our most precious resource–our children."

— Mark D. Hinderliter, Ph.D., Principal, Third Way, Inc.

"A cogent, informative book. The information was presented in an easy to read format, and the end-of-chapter summaries made focusing on salient points a breeze."

— Jane Webb, retired nurse and mother of two daughters and a son

“Explaining how the brain works is a fantastic way for teens to be mindful about their thoughts and feelings. The way Dr. Coates tackles this is brilliant. I’ve studied much of the research on teens, and I’m thrilled to have this book. It says to teens and their parents, ‘what is happening to you is hard and is not your fault; there is a lot of growth happening inside of you; welcome it with open arms.’ It takes parents and teens out of the blame game so they can think clearly about what to do.”

— Ronit Baras, parent coach and author of *Be Special, Be Yourself for Teenagers*

“I loved the book. It’s perfect timing for me, as I have a 12-year-old daughter! I loved the real-life examples. I was pleasantly surprised to discover that the actions are very doable, not overwhelming at all, and I’m already implementing some! Having the science and rationale behind it has made it easier to explain to my daughter how we can get her brain working better and what it will mean! She is now quite interested in expanding her way of thinking.”

— Cath Hakanson, author of *Girl Puberty, Boy Puberty* and *The Sex Education Answer Book*

“This is a very well-researched and logical book that provides clear evidence of what one must do to protect the PFC and why the formative adolescent years are so critical. I think a great place for it would be in health classes in schools, with an opportunity for kids to discuss among themselves in groups, guided by their teacher. Other places besides schools are churches, for pastors to help parents.”

— N. Elizabeth Fried, Ph.D., President, The Learning Engine, and award-winning executive coach

“Dr. Coates has provided a simple road map for any teen and parent to follow in order to take advantage of what’s been learned about how young brains develop. As he points out, never before in history have we had access to such powerful information. This short, easy to understand book enables parents and teens to take full advantage of this knowledge. A young person’s future success may depend on what is learned here.”

— Rick Stamm, MPsSc, founder, The TEAM Approach, Inc.

“As Dr. Coates tells us in *How Your Teen Can Grow a Smarter Brain*, teens have the power to improve their creative, critical thinking, and reasoning skills. This is great news for parents, teachers and anyone else who cares about our kids and their futures. We can help our teenagers get on track for happier, more successful lives, and Coates offers many specific tips on how to do that. The insightful chapter on screen time illuminates one of the greatest challenges of our time. As a parent of teenagers and a teacher of middle school kids, I heartily endorse this book."

— Mike Ferry, author of *Teaching Happiness and Innovation* and director of the *Creativity Retreat In Ireland*

"I so enjoyed this book! My boys are now in their early 20s, and we all would have greatly benefitted from this book for parents and teens. I especially liked the 'typical questions' suggested for parents to help teens to think on their own. This leads to the development of critical thinking, a necessity for any rational adult. The concepts of setting goals, foreseeing consequences and staying focused are incredibly important for their future in work and life in general. What a wonderful idea to have chapters for the teens themselves to read and gain deeper understanding of what they can do for themselves. All the parental influence in the world won't make teens become successful adults if they don't learn that 'no one can do it for you.'"

— Amielynne Barr, MHA, PHR

I wish I had this book when I was raising my children. From the beginning descriptions of the brain with "blossoming" and "pruning" to Chapter 5 with the questions parents can use to facilitate development of their children, all the way to the conclusion, this is a great read for parents as their children reach even the pre-teen years. Chapter 6 regarding screen time usage is so relevant today it's scary. I also love the chapters on what teens can do to help develop their brain and preparing for adolescence. This is a wonderful tool for parents.

— Nicholas J. Scalzo, Ed.D., CEO, OnTrack Training

How Your Teen Can Grow a Smarter Brain

Other Parenting Books by Dennis Coates

Conversations with the Wise Uncle

Conversations with the Wise Aunt

How Your Teen Can Grow a Smarter Brain

6 Game-Changers That Will Maximize Your Teen's Brainpower—Permanently

Dennis E. Coates, PhD

First Summit Publishing

How Your Teen Can Grow a Smarter Brain: 6 Game-Changers That Will Maximize Your Teen's Brainpower—Permanently

Printed in the United States of America
First Summit Publishing
An imprint of Performance Support Systems, Inc.
757-873-3700

Illustrations licensed for use in this book:
Cover: 3d render of a medical image of a male figure with brain highlighted, © freepik.com, KJpargeter
Chapter 1: Brain Anatomy—Frontal lobe, © canstockphoto.com, Decade 3D
Chapter 2: Neurons, © canstockphoto.com, iDesign
Chapter 2: 3d Model of House, © canstockphoto.com, 3000ad
Chapter 2: Small House, © canstockphoto.com, vencavolrab

Cover and book design: Paula Schlauch
ISBN: 978-0-692-19160-6

Dedication

For all the conscious, caring parents who, out of love, will do almost anything to help their children become happy, intelligent, independent adults.

Acknowledgements

I want to thank my wife, Kathleen Scott, and my business partners, Meredith Bell and Paula Schlauch for their constant encouragement and support during the past five years. Their enthusiasm helped me stay focused, and their feedback during the writing helped me achieve my vision for parents of teens.

The book in its final form was influenced by several others, who took the time to read this book and provide helpful feedback: Michael Gurian, Victoria Dunckley, Deborah Gilboa, Cath Hakansan, Elizabeth Fried, Amielynne Barr, Jane Webb, Mark Hinderliter, Mike Ferry, Rick Stamm, and Ronit Baras.

I'm also indebted to several people who generously shared their "teen journey stories" with me, which revealed the influences that shaped them as adults: Jack Pryor, Clinton Whitehead, Jordan Lovinger, Joe Paul, Eileen McDermott, Tucker Toler, Ron Wilson, Beth Westmark, Brooke Gunning, and Bob Hodges.

Finally, this book would never have been written without the inspired research of Dr. Jay Giedd and his team at the National Institute of Mental Health, which has revolutionized our understanding of adolescent brain development.

Table of Contents

Preface

"Where there's a will, there's a way" is a famous adage dating back centuries. It means that when a person finds the will to do something, that thing will motivate them to find the way to make it happen.

In practicality, I have more often observed the reverse. That is, "Where there's a way, there's a will," which means that if you give someone a way to do something that is "doable by them" and it leads to a desired goal, they will spontaneously summon the will to do it.

Your Teen Can Grow a Smarter Brain is that doable way that will give parents, teachers, counselors and teenagers and young adults the will to grow that smarter brain.

Because Dr. Coates delves into brain research and neuroscience and writes about it in such an easy-to-understand-and-follow style, any and all readers will discover a roadmap they can follow that, if followed, automatically results in developing a smarter brain.

This book does it by bringing a "sense of coherence" to the subject of learning and parenting *and* self-help (if a teenager wants to use it by themselves). A sense of coherence means taking something that was previously unmanageable or confusing and turning it into something that is comprehensible, understandable and meaningful. Taken together, "comprehensible, understandable and meaningful" turn what had been confusing and unmanageable into something clear and manageable.

When that happens, people spontaneously become empowered, emboldened and enthused about putting it into action. And when that happens, they begin to see results almost immediately.

That is what all readers will discover and do with this marvelous book.

— Mark Goulston, MD, psychiatrist and author of *Just Listen*

Introduction

I recently had a conversation with a local physician. Happily, we didn't talk about my health. Instead, he spoke enthusiastically about his three children, two boys and a girl, all of whom have left home. He told me something unbelievable about them: all three achieved a maximum score on their SAT and earned a National Merit Scholarship. The boys have gone on to successful professional careers, and his daughter, the youngest, recently graduated from an Ivy League school with *a quadruple major*. While he and his wife encouraged learning and education, they never talked about getting a college degree; *they only talked about attending graduate school*. The family had a tradition of eating dinner together, a time for discussing serious issues. He told me that as his kids got older they often challenged his opinions.

It's an amazing story, and of course it represents some kind of best-case scenario. Without being aware of the latest research about adolescent brain development, he and his wife did the No. 1 thing any parent can do to develop a massive foundation for critical thinking skills in their children's growing prefrontal cortex: they encouraged them to think. His story was a vivid example of how brilliant young adults have been nurtured throughout human history: doing the right things without knowing how adolescent brain development works. Those three kids were amazingly fortunate.

By contrast, for thousands of years, many other parents, without knowing it, failed to provide this kind of parental encouragement. As I'll describe later in this book, my own kind and responsible parents were among this group.

And of course, untold numbers of young people disrupted their own adolescent brain development by abusing alcohol or drugs. Again, without knowing they were doing so. Now that we have the facts about adolescent brain development, we appreciate that since the beginning of human history, this important aspect of growing up happened by chance, not by conscious effort.

Years ago at a family reunion, two encounters with very different 22-year-old men left me wondering.

Kyle seemed like a loner in the crowd. He didn't relate to the younger kids, and he seemed uncomfortable mingling with adults. I saw him sitting on a folding chair, staring off into the afternoon sky, so I sat down next to him.

After we talked awhile, I was struck with an overall impression that Kyle was stuck in adolescence. He had dropped out of community college and was living at home. He occasionally worked part-time, but he spent most of his days playing video games with his pals. I asked him, "If you could make money doing it, what would you like to do most?"

His reply: "Windsurf."

"Can you make money doing that? Like go pro or something?"

"No, I'm not that good. Not many people make money windsurfing. Besides, my parents wouldn't sponsor me."

I later met the other young man, Kyle's cousin, while dropping off a package at his home. Tom waved from the front door and came down the steps to meet me. "Here, let me take that. I have a place for it."

Tom's energy, confidence and strong self-esteem contrasted with my impression of Kyle. I was wearing one of my Duke basketball t-shirts, and I knew Tom was a graduate of Duke's arch-rival, UNC. So I grinned and said, "I hope you don't mind. I went to school there and I'm a big fan."

"That's no problem," he said. "I've applied to the Duke graduate school. I want to get into their psychology doctoral program." Clearly, he had a vision for his future and had set goals for himself.

The differences between the two young men were remarkable. One displayed significantly greater intellectual capacity. What influences in the two young men's youth had caused them to turn out the way they did? What activities had they been involved in? What courses had they excelled in? Did they have adult mentors who coached them?

From my understanding of child brain development, I knew the differences had nothing to do with their genes. Even though both had well-educated, successful parents, they developed different levels of brainpower unconsciously, without realizing how these skills are acquired.

Over the years, similar encounters with teenagers have motivated me to find out why some young people grow up prepared to tackle life's challenges, while others struggle to focus or make good choices.

Every parent I've ever met hopes their teen will end up like Tom, even though they've heard stories of the opposite. But realistically, as Tom's example illustrates, wonderful things can happen during the second dozen years of growing up. For example, a teen can:

- Become an avid reader
- Take studies seriously and become a knowledgeable, well-informed individual
- Learn how to learn
- Build executive function and critical thinking skills
- Get involved in service projects
- Become passionately interested in things that could lead to a career
- Set goals for the future
- Get a job and save money for college
- Develop a strong work ethic
- Maintain healthy nutrition and physical fitness habits
- Spend time out in nature
- Become responsible and independent
- Learn a lot of practical life skills
- Earn strong self-esteem and self-confidence
- Develop empathy, compassion, and sensitivity
- Improve relationship and leadership skills
- Make friendships that last a lifetime
- Have a lot of good, wholesome fun
- Strengthen bonds with parents and siblings
- Choose their own path of spirituality
- Get accepted by a branch of the military, a good trade school, or a college, prepared to do next-level work

These and other possibilities happen to young people all the time. Of course, teens make their own choices; parents can't force these outcomes to happen. But love, communication, guidance, support, and encouragement make a big difference.

Alternatively, your teen might:

- Use alcohol or drugs, which can disturb normal brain development
- Become addicted to smoking or other substances
- Get hooked on action shooter video games
- Spend too much time social networking
- Become fascinated with pornography
- Bully other kids or be bullied
- Break the law
- Get injured or die by automobile accident
- Be victimized by sexual predators or sex trafficking
- Get pregnant or cause a girl to get pregnant
- Contract sexually transmitted diseases
- Become overly anxious or stressed
- Suffer depression or commit suicide
- Become the victim of an eating disorder
- Inflict self-harm
- React to parenting with rebellion and defiance
- Acquire poor health habits that lead to obesity
- Lose interest in studies
- Feel dependent or entitled
- Suffer low self-esteem
- Submit to peer pressure
- Fail to discover purpose or ambition
- Fail to launch after high school

Parents can't lead their children's lives for them. As kids become more independent, they spend more time away from home than with family. Along their journey of growing up, they make choices. Most of the wonderful things and most of the awful things they do depend on their developing good judgment, which is a work in progress. It can be a jolt to remember what kids don't know at this age.

Will they become smart enough to use good judgment? The stakes are high:

- **Right now**, the ability to make wise choices will determine how a teenager negotiates the tricky path past the awful outcomes and towards the wonderful ones.

Will your child make it through adolescence without getting into serious trouble?

- **Later in life**, executive function and critical thinking skills are critical for success in careers, relationships, and life in general.

Will your child become the kind of adult who thrives while dealing with life's challenges?

And during the intervening years, through all the conflicts, heated moments, and hard lessons, will your relationship with your child survive and grow stronger?

Questions like these can keep concerned parents awake at night. You and your child could use a little luck.

Good luck is always welcome, but for nearly every young person, luck isn't the best solution. Adolescents need to understand what's happening in their developing brains and get skilled at solving problems, using good judgment, and making smart choices—thinking skills no one is born with.

What you'll discover in this book is that *for the first time in human history*, we know how young people can *intentionally* wire their prefrontal cortex (PFC) for dozens of powerful thinking skills, building the kind of brainpower that will allow them to meet challenges and assure success in their career.

Wow. Maybe you should read that paragraph again. A statement like that seems too good to be true.

But it's not. Throughout the ages there have always been exceptional people who arrived at full maturity with impressive intellects and went on to achieve extraordinary things. In their youth, they may have had mentors who stimulated their development. Or maybe they got involved in activities that caused extensive development in the smart part of their brain.

Some kids got lucky. At the same time, many others did not.

By contrast, an exciting promise runs through this book:

You can take luck out of the equation.

Thanks to recent research into the growing adolescent brain, we now know how this aspect of brain development works, so an informed and motivated adolescent can consciously and deliberately do the things that will make the best-case scenario happen.

Twenty years ago, a team of scientists were discovering evidence for what we now call "the teen brain." In the mid-1990s, brain researchers at the National Institute of Mental Health (NIMH), led by Dr. Jay Giedd, studied the brains of teenagers by using non-invasive functional magnetic resonance imagery (fMRI). It was previously thought that by the age of ten, children had completed all phases of basic brain development and that they were now building on these foundations. However, the new research proved this wasn't the case at all; young brains continue basic development throughout adolescence—especially in the prefrontal cortex (PFC), the "smart part of the brain."

The game-changing discovery that the PFC is still "under construction" throughout adolescence was welcomed with surprise and gratitude. Over a dozen books and countless videos, articles, and blog posts popularized this information. Today, many of the adults I talk to have heard about the teen brain.

This initial rush of information had a positive upside. Savvy parents learned to appreciate how much of the emotion, poor judgment and risk-taking characteristic of the teen years happens because the PFC, the executive functioning and critical thinking part of the brain, hasn't completed basic development. So instead of reacting to problem teen behavior with shock, anger or criticism, parents can choose to exercise empathy.

Unfortunately, while this new information has been useful, it often led to misunderstandings. For example, many caring adults assumed that the "craziness" created by the still-developing teen brain is a phase kids will outgrow by the time they're adults.

This is not always the case. As you'll learn in Chapter 2, while some young people acquire significant intellectual capacity along with their maturity, too many others do not. Furthermore, the dysfunctional behaviors of youth can actually become ingrained patterns they carry into adulthood.

Also, some parents have excused their teen's inappropriate behavior, believing the "growing pains" were not their fault. They gave their kids a free pass, when they should have encouraged them to try harder.

Now, with this book, parents can learn what it takes to develop the adolescent PFC and what they and their teens can do to deliberately maximize the result.

In a way, becoming strong intellectually is like becoming strong physically. A young person can adopt a routine for "working out" to build physical attributes such as strength, speed, and endurance. This isn't something you can do for your teen. They have to show up and do the work to get the results they seek.

Building strong mental capacity works the same way. If they "get the reps" by exercising thinking skills regularly during the teen years, the brain circuits that enable these skills will form and grow stronger. Like physical fitness, if they don't make the effort, not much will happen.

Sharing this information with your adolescent is essential to capitalizing on this unprecedented opportunity. While your teen will have to do the work, you can help immensely by giving information, support and encouragement. The goal is to help children bring an advanced intellectual skill set to their adult lives, one they can apply in challenging careers, lives and relationships.

My passionate interest in how the human brain develops, learns, and thinks began over 30 years ago, during the final years of my career as an officer in the U.S. Army, where I specialized in training and development. During those years I wrote training manuals and developed leadership courses for young leaders, mid-career officers and ROTC cadets.

After I retired in 1987, I co-founded Performance Support Systems, which creates computer-based adult learning systems. From the beginning I didn't want our users to simply love our products; I wanted them to acquire improved skills they would use for the rest of their lives. To achieve this, I needed to understand how learning actually happens in the brain. This inspired me to study brain research, an intense learning journey that persists to this day.

As I studied the research and then read what was being published about the teen brain, I realized that the most important conclusions from this research had yet to be explained: the long-term consequences of this final phase of brain development and how to maximize the result, which can vary from child to child, depending on how much thinking young people actually exercise during their teen years.

The stakes are high.

While a person doesn't have to be an elite thinker to find a niche and have a good life, teens who work on growing their intellectual capacity will set themselves up to deal with practically any challenge later in life. And they no longer have to leave it to chance.

This book is unique in a couple of ways. First, I've written it for parents, but I encourage them to pass the book along to their teenager, because not much will happen if the young person doesn't understand what to do or why it's important. In fact, I address my writing directly to the teen in the chapters of Part Three. However, I recommend that both parents and their teens read the entire book.

Throughout the book I mostly use the term "parent," but the information is just as useful for all adults who help teens prepare for life:

- Teachers
- Mentors
- Coaches
- Counselors
- Family members

Also, I often use the words "teenager," "teen," and "adolescent child" interchangeably, because the insights apply to all young people from age 12 to 24.

Only the young people whose brains are developing can do the work to wire their PFC, so in the best case you'll share the book with your child, whether a preteen, teen, or young adult.

Therefore I created a book a teen would read. I've avoided academic and technical language, and the chapters are brief. I included very few allusions to research and case histories—common elements in self-help books.

There's a lot going on in the adolescent brain, and descriptions of it can be daunting. So to drill down to what teens and their parents need to do, I've focused exclusively on the developing prefrontal cortex, not other brain areas that are still maturing. I invite readers who are interested in the complete picture of the still-developing teen brain to read neuroscientist Dr. Frances Jensen's excellent book, *The Teenage Brain*.

My goal is to focus simply on what parents (and other caring adults) and teens (adolescents) need to know and do to develop a robust intellect.

The book is divided into three parts. The purpose of Part One is to explain in the clearest possible terms what the prefrontal cortex will do if it is wired for critical thinking skills during adolescence. It also explains how the wiring happens. Hopefully, this new information will inspire a strong desire to do the work.

Part Two outlines how parents (and all caring adults) can support and encourage young people to explore activities that exercise the smart part of the brain.

While in the best case teenagers will read the whole book, I wrote Part Three especially for them. It tells teens what they need to do, including protecting their brain from substances that can disrupt brain development. On the plus side, they can get involved in interesting courses, extracurricular activities, and fun games that exercise the PFC.

It's possible that your teenager has already been involved in some of the activities I recommend, maybe for years! In this wonderful situation, this book will serve to explain and confirm the wisdom of what he or she has been doing. And it could provide insights and motivation for expanding these efforts, especially when facing new challenges and opportunities beyond high school.

Building a fine mind doesn't have to add more activities to an adolescent's already full plate. In most cases, doing the work simply means doing different things or doing the same things differently. It will mean making informed choices for the right reasons, all of which are fun, safe, and fulfilling.

Part One

WHAT YOU NEED TO KNOW

During the second dozen years of growing up, the prefrontal cortex (PFC)—the smart part of the brain—is ready for the wiring of basic skills for executive function and critical thinking. This is a once-in-a-lifetime opportunity to build the foundation for thinking skills needed to deal with life's challenges and become a successful adult.

The method is simple: exercise a variety of thinking skills repeatedly throughout adolescence. While this effort can happen by chance, this book empowers a young person to do the work by *choice*. Parts Two and Three describe a number of ways to do this.

But first, in order to gain the confidence and motivation to make these choices, it will help if you understand what the payoff is and how thinking skills get established.

This is the purpose of Part One:

Chapter 1: The prefrontal cortex can be wired for a myriad of foundation thinking skills. Understanding what these skills are can create a strong incentive to do the work.

Chapter 2: How the brain gets wired for these skills is both fascinating and simple. Like acquiring any skill, to make it their own, the teen will have to do the reps. But the clock is ticking: whether or not they establish the brain circuits for powerful thinking skills, the PFC is steadily eliminating the unused brain cells and connections. By the end of adolescence, only the neural pathways that were wired will remain.

Chapter 3: The big ah-ha is that with this knowledge teens can literally take luck out of the equation. Parents don't have to just trust that teachers will stimulate their child to think or hope that they will somehow grow up smart enough to deal with life's challenges. A teen can intentionally and consciously do what it takes to grow a smarter brain.

1

The PFC is the Smart Part of the Brain

"The frontal lobe is often called the CEO, or the executive of the brain. It's involved in things like planning and strategizing and organizing, initiating attention and stopping and starting and shifting attention. It's a part of the brain that most separates man from beast, if you will. That is the part of the brain that has changed most in our human evolution, and a part of the brain that allows us to conduct philosophy and to think about thinking and to think about our place in the universe."

– Jay Giedd, MD –
Pioneering adolescent brain researcher, NIMH

This chapter describes the impressive functions of what I often call "the smart part of the brain"—the prefrontal cortex (PFC).

To be accurate, the thinking skills facilitated by the PFC aren't the only aspects of intelligence. In his book, *Frames of Mind* (1983), Howard Gardner proposes several kinds of intelligence. For example, visual-spatial intelligence is the ability to understand what is seen. Verbal-linguistic intelligence is one's facility with words and languages. Bodily-kinesthetic intelligence is about physical control and coordination. Musical-rhythmic intelligence has to do with sounds, tones, and music. Each of these types of intelligence is enabled by extensive programming in specific areas of the brain.

Interestingly, several of the other kinds of intelligences he describes, such as interpersonal, intrapersonal, naturalistic, existential, and logical-mathematical, are heavily influenced by the PFC.

The PFC is the relatively large outer area of the brain located right behind the forehead. It receives input from other areas of the brain to "connect the dots," relating emotions, perceptions, memories, concepts and thoughts. For example, it can relate a sight or a sound to an emotion, which can establish the importance of what's happening. It can compare past memories with present events to discover cause-and-effect relationships. And based on this understanding, it can form images of the future. And dozens of other mental tasks that help you get through life successfully.

This ability to make sense of things—and to remember what was learned—is why I call the PFC the smart part of the brain. Most mammals have a functioning PFC, but their tiny PFCs are far more limited than that of a human being. If you've lived with dogs or cats, or if you've observed squirrels or raccoons, then you know they show signs of intelligence. But their capacity for

The PFC is the Smart Part of the Brain

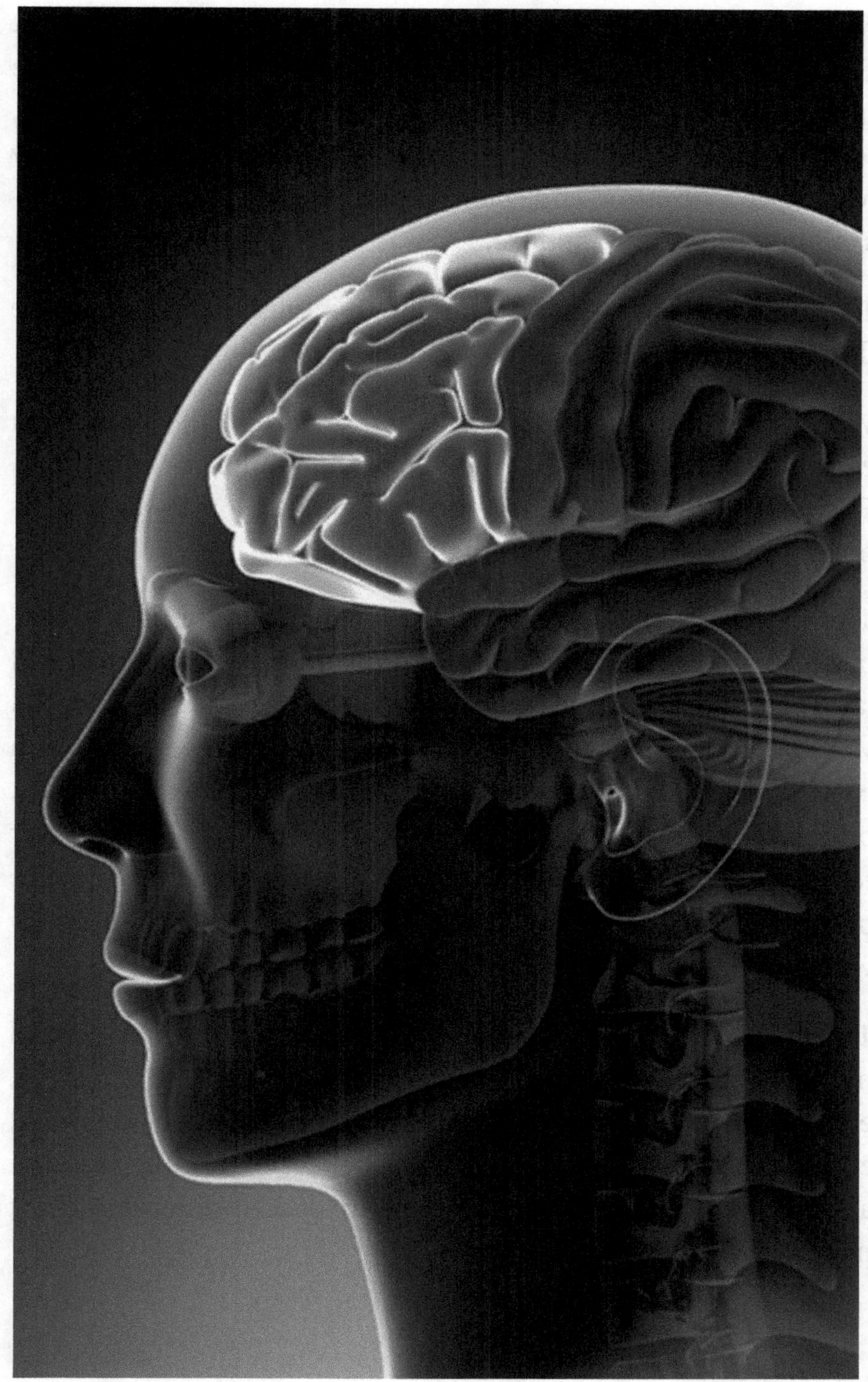

reasoning, decision-making, or self-management is primitive. Consequently, they rarely consider the possibilities before taking action. They rely mostly on *stimulus-response* reactions, while humans (thanks to their impressive PFC) have the ability to analyze a situation and consider alternatives before they act—*stimulus-analysis-response*.

This capability is a game-changer. It's why humans are the dominant species on Earth—not dogs or cats.

When I refer to the PFC as the smart part of the brain, I'm certainly not implying that the rest of the brain isn't important. Each area helps make us effective in other ways, such as perceptually, physically, or linguistically.

Throughout this book I refer to the "foundation thinking skills" of the PFC, because the thinking skills developed in youth will later serve as the basis or foundation for more advanced versions of these skills, such as diagnosing an illness, arguing a case in court, organizing a project to build a highway overpass, or deciding how to market a new kind of product. For an adult to acquire specialized abilities such as these, the circuits for basic thinking skills need to be in place, and the opportunity window for this foundation development is adolescence.

The mental skills that operate in the PFC are known by several names: social-emotional learning, executive functioning, emotional intelligence, critical thinking, creative thinking, conceptual thinking, intuition, and intellect. To appreciate what a thoroughly wired PFC can do, here are descriptions of its most impressive functions:

Understanding

As we learn how things relate to each other and why things happen, we make sense of our world. We form concepts, theories, principles, conclusions, assumptions, attitudes, and opinions. We become aware of what things mean and how things work, and we seek to learn more. We question assumptions and consider different points of view. To create understanding, the PFC helps us determine:

- How things are similar or different
- Assumptions, theories or principles of cause and effect
- Which factors contributed to an outcome
- How a device works
- What a situation means
- What is communicated by a facial expression
- How important something is to you
- What people need and want
- Why someone behaved a certain way
- The emotions or feelings behind someone's words or actions
- What an artist was trying to express

For example, you can look up at the sky and simply notice that it's a partly cloudy day, and think: "The clouds are moving really fast today." Or you could identify what kind of clouds you're seeing: cirrus, cumulus, cumulonimbus, or other types of clouds and the kind of weather associated with them.

Some kinds of knowledge are actually misunderstandings of the way things are. For example, lots of people think that because they can walk and chew gum while reading a map, they can also safely multi-task while driving. In fact, driving a car at 50 miles per hour requires the ability to react to danger in less than a second. What most people don't know is that an area of the brain called the thalamus receives input from the senses and tells the rest of the brain what to focus on. This inability to focus on all aspects of input simultaneously is a good thing: it makes effective thinking possible. Multi-tasking actually involves shifting attention back and forth. So when you look at your cell phone, even for a second, your attention is shifted away from the road ahead during that time. Understanding this fact can motivate you to wait until you're stopped before checking your phone—and possibly save your life.

In a lifetime, you can come to understand millions of things. It begins with parents' explanations, continues throughout one's formal education, and can lead to a lifelong journey of learning. Or not. I've met many people who are quite happy with what they know and ceased to consider new facts or knowledge a long time ago.

Foreseeing consequences

One way of relating one thing to another is to observe how one event follows another. Does one event cause the other? If so, then later when we observe the first event again, we can predict the outcome. The more we make these kinds of associations, the better we can imagine future consequences, set goals, make plans, and manage our lives. The advantages of this mental skill are endless. Our PFC helps us think about future consequences:

- What is likely to happen in this situation
- What you can do to create a certain outcome
- The future benefits to a course of action
- The downsides to a course of action
- The costs of a course of action
- Why doing something doesn't feel right
- The most effective sequence of actions
- The best way to get what you want
- The best first step to achieve something

Often, when a small child wants something and doesn't get it, the frustration will lead to tears. If the parent gives the child what she wants in order to stop the crying, the child may conclude that crying is an effective way to get what she wants—an unfortunate, unintended learning.

My wife loves gardening. One summer, a plant she was caring for began to wither. It was puzzling because she had prepared the soil and watered it regularly. She had not foreseen that this plant needed full shade to thrive, and it was getting too much sun each day. So she moved the plant to a shady area, and it began to recover.

Learning from experience

It is said that experience is the best teacher. But even though lots of things happen to us, it's possible to go from event to event or from one mistake to another and fail to learn anything. But if a person analyzes an experience by asking thoughtful questions about it, one can understand why something happened and be better prepared to handle similar situations in the future. You can consider:

- What happened
- Who did what
- Who said what
- Why the event happened
- The consequences of what happened
- The costs and benefits
- What could have been done differently to achieve a better result

The U.S. Army has formalized this thought process. After a combat mission, the leader holds a meeting called the "after-action review" to identify why things happened the way they did and what could be done to achieve a better result.

Albert Einstein is often credited for saying, "Insanity is doing the same thing over and over and expecting different results." It's not true insanity, of course, but neither is it effective thinking. I once knew a fellow who was consistently late for work. Even though this repeated failure put his job in jeopardy, he didn't change the way he got himself up in the morning. He never reflected on why this was happening, the consequences, or what he could do differently. Consequently, his boss questioned his commitment to his job.

Evaluating

When we compare an action, a process, or a product to a standard of quality, we judge how good, valuable or effective it is. Concepts of quality are learned and stored in the PFC, and are used to create judgments about what's going on in the world. We can establish:

- How much something is worth
- How well something is made
- How admirable an achievement is
- The merits of an accomplishment
- The flaws in an accomplishment
- How well something does what it's supposed to do
- Which things are more valuable than others
- How safe something is
- How nutritious something is
- The total cost of something, compared to the benefits

Someday, you may walk into a furniture store wanting to buy a new chair. It will help if you can tell how well the chair you're considering is made. Is it the kind of chair that will survive many years of use? Can you tell whether it's worth the price the store is asking for it? And what about the quality and pricing of similar chairs in other stores?

The same questions will apply when you go to buy a car—or a house!—where evaluating quality and value is even more difficult. Sooner or later everyone gets burned by a bad buying decision. The habit of doing a little learning about quality and value can save you a lot of money over time.

Reasoning

It's not easy to be logical. Similar to playing chess, which requires thinking ahead to the next moves, a person can acquire an impressive array of reasoning skills. But how smart your PFC is depends in large part on how many reasoning skills you have. When you get involved in problem solving, listen to other points view, or take courses in science, logic or philosophy, you learn that there are careless or fallacious ways of thinking. In other words, you have to train your brain to be logical. In the absence of these special thinking skills, you might easily make false assumptions about the world. With practice, you can get better at:

- Determining whether a statement is true
- Deciding whether an assumption violates a proven principle
- Judging whether an argument makes sense
- Judging whether an opinion is consistent with known facts
- Determining the chance that what has happened could happen again

I once knew a woman who would buy ten lottery tickets twice a week. Her reasoning: buying ten tickets would give her ten times as many chances of winning; and since she always played the same numbers, she would be more likely to win with each passing week. She was disappointed to learn that her odds of winning were only 1 in 5 million for each ticket played, no matter how many times she played it. She stopped buying ten tickets each time, instead buying just one. Her reasoning for this action actually made sense: it's worth a dollar for the fun of thinking about the possibility of winning.

Learning about logic can help you make good choices. To persuade you to buy something, ad writers often deliberately exaggerate a claim or try to excite your emotions. Lazy thinkers fall prey to these appeals. Politicians often use similar tactics to get you to vote for them. They count on enough voters simply accepting what they say, if they say it with conviction.

Problem solving

In a typical day, you may be frustrated by mistakes or failures. Something goes wrong. After gaining weight, the clothes you want to wear to a party no longer fit. Your computer gets hacked. The lawn mower doesn't start. You can't afford the tuition to your first choice of colleges. Things wear out and break down. Maybe your situation changes and you aren't sure what to do. You might have an opportunity, and you aren't sure whether to pursue it. Or you want to improve a relationship. When you realize something is no longer working, you can:

- Find out what isn't working right
- Discover what's causing the troublesome indication
- Think about ways to correct it
- Recall whether you've seen this kind of problem before and how well your fix worked
- Find out how other people have handled the problem
- Decide whether taking a different approach altogether might achieve a better result
- Evaluate what other people suggest

- Consider the possibilities, along with their pros and cons

Creative thinking

Based on what you've read so far, you understand that your PFC can make sense of what is. It can also, if sufficiently developed, imagine what could be. It can instruct the visual cortex to construct images that are variations on what exists in present reality. It can combine images to produce an idea with the positive aspects of both. In other words, it can entertain what's possible. It can also link these images to practical considerations to determine how feasible they are. Make a habit of doing this kind of thinking, and you can:

- Think of better ways to do something
- Come up with new solutions to a problem
- Envision images that express your feelings and values
- Interpret the meaning of artistic expressions
- Imagine a possible future

When you're trying to solve a problem, it's helpful to consider if it's a "fix it" problem or a "change-it" problem. When something that usually works fine suddenly stops working, it's probably broken in some way. For example, the cold water dispenser in your refrigerator stops dispensing water. You have to "troubleshoot" to find out what part of the system failed. Maybe you need to adjust a setting or replace a part. If the fix is too expensive, you might decide to get a new refrigerator. This process involves logical thinking, not creative thinking.

Creative thinking is needed when an old solution no longer works and you need to try something different. To do that, you need to imagine what that something different could be. In most cases, there will be many possible approaches. Considering all these ideas is often referred to as "brainstorming." The trick is to let your imagination run free without judging the pros and cons, which you can do later.

Creative thinking is how engineers and inventors come up with improved ways of doing things. Of course creative thinking can help you do things differently and better in your everyday life. Years ago, my wife and I tried to figure out how to save money to buy things we couldn't afford. Often, we'd just live without them. Then we decided to consider an "outside-the-box" approach. Even though we liked where we were living, what if we moved to a smaller house? The house we found cost less than the one we sold, so we paid off the mortgage. And insurance, taxes, and monthly utilities cost less for the smaller house, which was more efficiently designed than the one we left. We ended up with an easier lifestyle and more money to buy things we wanted.

Decision-making

At the end of problem solving, the next step is action. What will you do to address your issue? Will you react emotionally? Instead of reacting, you can think about what to do. What choice will you make? Will you do what you've always done in this situation—fall back on your old habit? Will you let someone else decide for you? Or will you think consciously about what to do? If you believe there is more than one way to fix a problem, you can foresee and evaluate the risks, rewards, costs, and benefits of each option—a mental process that can only happen in the PFC:

- Consider how much time you have to think through your options, and whether you need to act right away
- Determine your options
- Compare the advantages and disadvantages of each option
- Evaluate whether any option has an unacceptable downside
- Ask others what they recommend
- Imagine how actions could affect people
- Get in touch with what your gut says—whether it feels right

A typical decision every young adult makes is what to do after high school. Whether it's college or training to be certified in a trade, there are likely to be many options, each with pros and cons, costs and benefits. Which university offers the programs that suit your goals? The same is true of going into military service; what are the advantages and disadvantages of each service? You'll need to do a lot of "due diligence" and consider a lot of factors.

Planning and organizing

Much of life is trying to get what you need and want: money, travel, a car, a career, and so on. The PFC can be wired for skills such as goal setting, planning a logical series of steps to achieve a goal, getting organized and managing the execution of a plan, and indeed, scrapping a plan in order to create a new one when life throws new challenges at you.

- Set a goal to get something you want
- Consider how this goal ranks with your other goals and choose which are more important to you
- Think about what you'll need to do to achieve them
- Decide the most effective sequence of actions, and what you'll need to do first
- Consider whether any of these actions conflict with other priorities
- Determine which resources you'll need and how to get them
- Think about who can help you if you need it

A friend of mine told me this story about his teen years. He had six older sisters, but none of them would let him borrow their car to go to a dance. So he walked into town. When he got there his boots were muddy, and he was so ashamed of his appearance that he turned around and walked home. He was so mad about this incident that he set a goal to buy his own car, even though he hadn't saved any money. He got three after-school jobs, and in six months he had enough money to buy a used car. A side-benefit to having his own car was he not only had found a way to earn steady money, the car gave him his independence.

British author Lewis Carroll famously said: "If you don't know where you are going, any road will get you there." But once you know what you want, you can imagine what you'll have to do to get it—a series of actions called a *plan*.

Attention and focus

Because the PFC is heavily interconnected to other parts of the brain, it receives real-time input, and it can direct the activities of other areas of the brain. For example, when you dream, remember or imagine, the PFC is sending signals to the visual cortex to create images. If the amygdala, which reacts to new stimuli with fight-or-flight alarm signals, tries to trigger behavior, the PFC can quickly evaluate sensory input and decide whether to tell the amygdala to calm down. If the senses notice something, the PFC can evaluate how important the information is, and then direct the brain to focus attention on it. In short, the PFC can help you stay focused while you work towards your goal:

- Evaluate how important it is to keep on doing what you're doing right now
- Identify your current top priority
- Shift your attention from one input to another
- Decide what you want to accomplish in the next hour
- Say "no" or "later" to appealing distractions
- When someone wants you to do something else, say "no" in a positive, considerate way
- If you get distracted anyway, think about what you can do to get back on track

I've seen this hundreds of times: someone is working on a project and a friend will call or show up for friendly conversation. The person's work comes to a halt while the two of them talk. Sometimes they will go somewhere, and no more work is done on the project that day. It's too easy to take a break, to shift from doing something that requires concentration to doing something fun.

There's nothing wrong with taking a break, but upon return, it takes a while to get your mind back in the flow of the project. This can happen even if the project is both urgent and important. The solution is say "no" to the distraction in a polite and considerate way: "Hey, I'd really like to do that. But I can't right now. I absolutely have to get this done. Rain check?"

Self-regulation and impulse control

When you get angry, you might express rage by striking out or saying something hurtful. While this is a common reaction, it's possible to be aware of your emotions, consciously pause until the emotions subside, and consider more effective responses. These mental activities take place in the PFC. And yes, teenagers, who may react emotionally while their PFC is under construction, can practice these skills, too. If they don't, they may act irrationally and even grow up to be adults who have trouble controlling their emotions. Self-regulation and impulse control involve:

- Being aware of what you're feeling
- Analyzing why you feel this way
- Imagining what could happen if you do what you feel like doing
- Deciding to wait before saying or doing something
- Thinking about what you can do to calm yourself
- Considering the most effective thing you can say or do right now

When I was a young captain in the Army, I served a combat tour in Vietnam. I commanded a mobile advisory team, and during that year I was involved in more than 100 combat missions. The whole point of these missions was to find and make contact with the enemy. Doing my job and surviving depended on my ability to "stay cool" while under fire.

As a result of all these experiences, I wired my PFC with a skill I call "mental toughness." For the rest of my life, when faced with a crisis, I would allow myself to feel emotions such as alarm, frustration, or anger, but I wouldn't do or say anything until the emotions subsided and I had a chance to think this thought: "What's the most effective thing I can do right now?"

The foundation thinking skills facilitated by the PFC are powerful.

In my opinion, calling the PFC "the smart part of the brain" isn't a clever exaggeration. If anything, it understates the PFC's true capacity. Basic thinking skills can be learned, practiced, and mastered. But this aspect of intellectual development needs to happen during adolescence, and it won't happen just because a young person is growing up. Only the growing adult-in-progress can do the work to wire his or her brain.

For thousands of years, all this circuit-building happened not by design or intention, but by chance. For a fortunate few, something grabbed their attention and caused them to think a lot. Some kids became passionately interested in areas of learning that affect PFC development. Some kids had wise parents or mentors who stimulated them to think for themselves, thereby exercising and wiring the PFC.

Others weren't so lucky. For them, adult life without many of these skills was more difficult. This is one of the reasons why people have always had noticeably different intellectual capacities.

If given an awareness of the amazing power of the PFC, will your child feel it's worth the effort to exercise these skills while growing up?

What it takes to intentionally wire the PFC—how skill development actually works—is the subject of the next chapter.

Major takeaways from Chapter 1:

- The PFC is the part of the brain that handles important executive and critical thinking functions.
- The foundation circuits for each of these basic thinking skills needs to be wired during adolescence.
- For thousands of years until the 21st century, the young people who did the work and created these forms of intellectual capacity did so without knowing that they were doing so.
- An awareness of PFC development means a teen can create brainpower consciously and deliberately.

Recommended actions:

- For the parent: As you read the next chapter, find out what a young person will need to do to wire their PFC for the skills mentioned in this chapter.
- For the teen: Think about how important each aspect of PFC intellectual capacity is to you. How would the

thinking skill help you at this time of your life? How would the skill help you achieve your future life goals?

2

Growing a Smarter PFC—How This Works

"[Blossoming] is a process we knew happened in the womb, maybe even the first 18 months of life. But it was only when we began scanning the brains of children at two-year intervals that we detected a second wave of over-production. And this second wave of overproduction is manifested by an actual thickening in the grey matter, the thinking part, in the front parts of the brain."

– Jay Giedd, MD –
Pioneering adolescent brain researcher, NIMH

Knowing how powerful a growing prefrontal cortex (PFC) can become, a young person may wonder, "How does this work? Is there a way for me to make my brain smarter?"

The answers are fairly simple—and interesting. The process for wiring a teenager's PFC is the same as the process that developed other areas of the brain earlier in childhood.

How It Works

Think of the brain as a computer made of human tissue. A working brain, like a computer, consists of hardware, software, and data. The brain itself is the hardware. Circuits of interconnected brain cells that make the brain function are like software programs. And to function, these programs process data. The data—the input to these circuits—come from the senses, other parts of the body, and stored images and concepts.

All behavior, whether physical or mental, is initiated by the brain. When you do something you've never done before, you have to concentrate to involve various brain cells that aren't yet connected. This is why at first the action feels difficult and awkward. However, if you repeat the behavior many times, the brain will be stimulated to gradually connect a physical network of brain cells that make it happen efficiently. These hard-wired programs in your brain are what enable skills, habits, and behavior patterns.

Like physical skills, mental skills don't develop simply because the time is right. Brain circuits don't wire unless you do the work to establish them. Even before you've acquired a skill, you can still concentrate on doing something new, but the action won't be easy or automatic. Think about what it takes to get

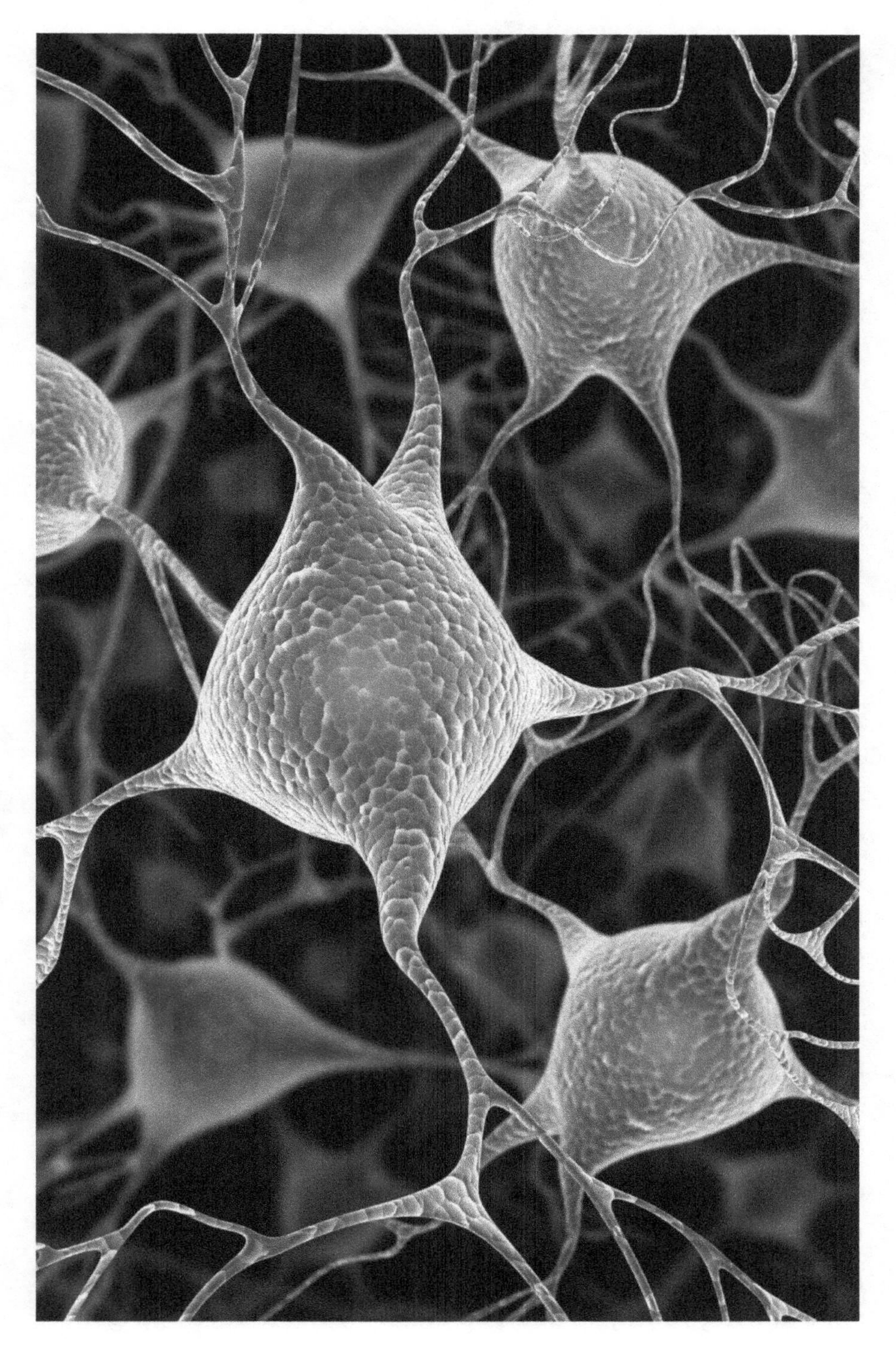

good at a sport or to learn to play a musical instrument. It takes lots of practice to stimulate brain cells to connect into circuits.

This is how a child develops the powerful mental skills managed by the PFC, described in Chapter 1. It's simple: they repeatedly perform the action, which stimulates the brain cells involved in the behavior until they eventually connect.

Foundation skills have to be wired first.

A person can learn to give a speech, debate a point of view, sing a song, act in a play, verbally translate a foreign language, or use any number of effective communication skills. All these abilities depend on having basic language skills, which are learned in early childhood. It takes quite a lot of effort for a small child to wire the language areas of the brain for foundation speaking skills. The advanced applications of speech can be learned later.

A one-year-old won't all of a sudden stand up and start tap-dancing. She has to develop foundation skills first: balancing, then standing, then walking, and then running. After these basic skills are wired, she can then focus on riding a bicycle or learning to tap-dance, if that's what she wants to do.

Thinking skills are no different. Just because a kid is now in middle school doesn't mean she can all of a sudden argue a legal case in court, draw up plans for a highway bridge, or diagnose a disease. She first has to develop a foundation of executive and critical thinking skills, such as self-control, focusing, understanding, reasoning, problem solving, decision-making, planning and organizing. If a person acquires these thinking skills during youth, an adult can later work on developing more sophisticated skills in areas such as law, engineering, or medicine.

The first phase for wiring a foundation skill is "blossoming."

When a child is born, the many areas of the brain have been prepared for development by an over-production of brain cell connector fibers called dendrites. This ensures there'll be enough brain cells and connections to construct all the basic skill circuits. Consequently, an infant's brain has many times more dendrites (connector fibers) than will ever be needed.

Once toddlers are walking around, they begin exercising the kind of cause-and-effect thinking they need to understand what's going on in the family's protected environment. This is why kids ask "why" a lot at this age. The more they learn, the more their young PFC wires itself for primitive thinking skills.

But the kind of mental capacity small children develop won't be enough to serve them when they're independent adults trying to make their way in the complex adult world. So at the onset of puberty, growth hormones stimulate *a second blossoming of dendrites* in the PFC.

For a teenager, this arrival of new connector fibers creates an opportunity to connect brain cells for powerful thinking skills. But at first, with so many new unused connections crowding the PFC, practicing critical thinking can be hard. To illustrate how disabling all these not-yet-connected fibers can be to a pre-teen, I'll use the analogy of road construction.

Where I live, the north side of town is in an area of rolling hills, and the south side of town is in a plain. So the main route that goes north through town is flat until it begins a steep ascent approximately in the middle. As the population grew, sections of this two-lane road were widened and improved. The easiest traffic areas—before and after the steep ascent—were constructed first. But the section that climbed the hill remained narrow and became a bottle-neck to traffic.

When the city undertook the complex task of shoring up the hillside in order to widen and improve the road that climbed the hill, it had to close this area to traffic for more than a year while they constructed a bridge and new traffic lanes. Predictably, the detours caused irritation, delays and a few accidents.

Something similar happens in an adolescent's PFC. Even though powerful new critical thinking pathways may be under construction, at first it's harder for a young person to use what we call "good judgment." There are too many unused connections to make efficient functioning possible.

So sometimes default thought processes redirect them to the amygdala, which is a primitive part of the brain used for emotional reactions. As a result, teens may act impulsively or take imprudent risks just for fun without thinking about the consequences. They may experience mood swings and surges of emotion and not know why. It's a difficult time of life, both for teenagers and their parents.

Now that brain science has explained what's going on, we can be more understanding of post-pubescent behavior. At the end of adolescence, after the PFC's construction is over, hopefully young adults will have wired this part of the brain with circuits for self-management, judgment, and critical thinking.

After blossoming comes "pruning."

While brain cells in the PFC are interconnecting, another vital process, called "pruning," is underway. Over-production of cell connections ensures that enough connections are available. The downside to having more connections than are needed is that the unused connections interfere with the brain cells that are trying to form circuits. So while certain brain cells are forming circuits, the brain slowly absorbs the nearby unused connector fibers. It may seem shocking that brain cells and connections are steadily being eliminated, but pruning prevents the possibility that electro-chemical impulses will be short-circuited.

Dr. Jay Giedd, pioneering adolescent brain researcher, summarizes it this way:

> *"How does the brain become the brain? It does so through two simple but powerful processes. The first is over-production. The brain produces way more cells and connections than can possibly survive....After this vast over-production, there is a fierce, competitive elimination, in which the brain cells and connections fight it out for survival. Only a small percentage of the cells and connections make it."*

At the same time, as circuits form and pruning does its work, an insulating substance called myelin surrounds each of the brain cells in a circuit, creating even more efficiency. As a result, the speed of electro-chemical impulses through the circuit increases dramatically. This is why a behavior pattern that once felt awkward will seem effortless after a skill circuit is fully developed. This physical reinforcement also makes the circuit permanent, which is why once you learn to walk or speak, you don't have to relearn these basic skills.

There's a limited "window of opportunity" to develop the PFC.

With brain cells and connections steadily being eliminated, a teen doesn't have forever to grow a smart brain. It's not something a teen can put off until later.

Again, Dr. Giedd:

"Those cells and connections that are used will survive and flourish. Those cells and connections that are not used will wither and die. So if a teen is doing music or sports or academics, those are the cells and connections that will be hardwired. If they're lying on the couch or playing video games or watching MTV, those are the cells and connections that are going to survive."

The process is like a sculptor making a work of art out of a block of marble. At the end of the artist's work all the unused marble will have been chipped away, and only the finished work will remain. As far as the PFC is concerned, after pruning has done its work, only the brain cells exercised over and over for basic executive and critical thinking will remain. Hopefully, at the end of adolescence, the young adult will have exercised many different thinking skills early and often so that the foundation that remains is extensive and robust.

A young person literally has to "use it or lose it."

Most people don't realize that the use-it-lose-it principle is at work during critical periods of early childhood when, for example, an infant is learning to see or a toddler is learning to talk. In the 1960s, researchers David Hubel and Torsten Wiesel did a controversial experiment on cats. They sewed one of a newborn kitten's eyes shut during the first three months—the sensitive time window for visual development. After the sutures were removed, the cats never recovered normal use of this eye. Adult cats treated the same way showed no visual impairment. They tried the same experiment on monkeys. When sutures were removed after six months, the animals had lost vision in the one eye. For their work, which showed the importance of critical periods of brain development, Hubel and Wiesel were awarded the 1981 Nobel Prize in medicine.

Related examples are the rare cases of "feral children," who had been lost or abandoned in early childhood and raised by animals for years, with no human contact during the critical period for language development. Later, when therapists tried to help them learn to speak normally, they were usually unsuccessful.

Some of these stories have little evidence to back them up, but one validated example is the 12-year-old "Wild Boy of Aveyron," who was found in the woods near Paris in 1800, acting like an animal. A school teacher adopted him and tried to socialize him. After several years, he could not teach the young man to talk.

A case closer to home is that of Danielle Lierow. When she was six years old, someone saw her gaunt, blank face in a broken window in Plant City, Florida, and called the police. They found her on a mattress covered with roaches, maggots and her own feces. She was like an infant: she couldn't talk, eat solid food, or play with toys. After six weeks in Tampa General Hospital recovering from malnutrition, doctors found nothing physically wrong with her. At the time of this writing she is 20 years old, but in spite of loving adoptive parents and years of efforts by therapists and teachers, she has not learned to talk. The sensitive window for learning speech had come and gone and the language areas of her brain had never developed; the cells and connections in the language areas of her brain had been pruned away. Her story has been documented in the Pulitzer Prize-winning article in the *Tampa Bay Times*, "The Girl in the Window," by Lane DeGregory, and her follow-up stories, "The Girl in the Real World" and "The Girl Ten Years Later."

The window for development opens, and after pruning has eliminated all the unused cells and connections, the window closes. The consequences are huge. *Use it or lose it.*

The more often a teen exercises the PFC, the more thinking skill circuits will form and fewer connections will be eliminated.

On the other hand, a teen who isn't giving his PFC a workout will end up with fewer skills and many more connections will be eliminated.

The resulting foundation—whatever it turns out to be—will be permanent.

Repeated exercising of many basic thinking skills during adolescence will establish an ample foundation, upon which an impressive edifice of applied thinking skills may be constructed.

If little work was done during adolescence, more learning is still possible. But the minimal foundation will not support the development of many applied thinking skills.

It's impossible for the eliminated connections to regrow. No do-overs.

On the other hand, adults who constructed a minimal foundation for intellectual capacity when they were young can still build on it if they're motivated to do so. This is what brain scientists call *neuroplasticity*, meaning the brain can change by wiring new circuits. But the new circuits will have to connect with the circuits that are already in place. The reality: you can continue to learn throughout life, but you won't be able to build a massive edifice on a tiny foundation.

Most teens aren't motivated to do the work.

It's instructive to compare the brain development of a small child with that of a teenager. All the self-programming a young child does is hard work. Think about it. How can you do the reps for crawling when you don't know how to crawl; to haul yourself up to a sitting position when you don't know how to sit; to stand with balance when you don't know how to stand; or to toddle when you don't know how to walk? As they say, it's like pulling yourself up by your own bootstraps.

The answer is that kids are highly motivated to keep trying, even after they fail a hundred times—until the brain circuits form the neural pathways that enable the basic skill. They're desperate to get what these skills will give them. They see stuff they want, they see how older family members get it, and they want it badly enough to repeatedly try and fail until the brain circuits form and the skill gets easier.

However, when it comes to learning thinking skills, teens lack the same motivation. It's not because they don't care about being smart. It's because, for a variety of reasons, thinking skills are quite different from the basic skills they acquired as small children.

Teens may know adults who think clearly, use good judgment, and make sound decisions. But they don't appreciate that these abilities, like any other skills, have to be developed through repetition. They don't understand that if they don't do the work, the skills won't develop. They may not even know they have a PFC, what this part of the brain is for, that these foundation skills are vital to their future success, that the years of adolescence are their only chance to do the work, or what they have to do to make it happen.

Not everyone ends up with the same intellectual capacity.

The next time you're out in your community, consider the people around you. Not everyone has the same intellectual "gifts." Not everybody has the brainpower to pursue higher education or intellectually challenging careers. Many of the people you encounter in life are bright. Others aren't. It may seem heartless to say so, but some people will never be able to work through the challenges of a complex profession. Some people will always struggle when dealing with the adversities of modern life and relationships, even as they try to make the most of what they have.

In other words, the "craziness" of teen behavior isn't a phase young people outgrow as they get older.

They will seem to "grow out of" some of it only to the extent that they exercise thinking skills before the sensitive window of PFC growth closes. At the end of adolescence, young adults who did the work will have ingrained robust circuitry for critical thinking. Others will have trouble connecting the dots—for the rest of their lives.

The stakes are high.

The all-important question: How much conscious effort will a teen put into wiring his or her PFC?

Learning about the amazing potential of the prefrontal cortex can help spark their motivation. And learning how the wiring process happens—what a young person has to do to establish a new skill—can help them understand why they must make an effort.

The good news: the second wave of blossoming and pruning in the PFC can be an extraordinary opportunity. In the past, only neuroscientists understood what the PFC does and how it develops. Kids needed some luck to do the things that help them grow up smart. Not knowing how certain school subjects and other activities exercise the PFC, they simply followed their instincts. Also, if they were fortunate, parents and mentors guided them to be curious, ask questions, and think.

With this new information about teen brain development, a kid doesn't have to get lucky. Knowledge is power. Young people can, by choice, learn what's happening and deliberately choose activities that will wire their PFC. With this information and the motivation to use it, they can intentionally make their brains smarter.

In the next chapter I share my own story about how, back in the day, my parents and I were ignorant about all this; but I was one of the kids who got lucky anyway.

Major takeaways from Chapter 2:

- Foundation skills need to be wired before a person can build on them with more advanced skills.
- Throughout childhood, the brain creates circuits for foundation skills, one functional area at a time.
- The PFC is the last area of the brain to be ready for basic development. This happens during the 10-12 years of adolescence.
- Building foundation thinking skills works the same way as building skills mastered in early childhood: if you repeat the action enough times, physical circuits that enable the skill form in the brain.
- The process for foundation skill development involves over-production of brain cells and connector fibers to make sure there are enough to support maximum learning.
- To make connected circuits more efficient, unused cells and connections are pruned away, and the cells in the remaining circuit are "insulated" with myelin.
- After all the unused cells and connections have been removed, a person can build on the foundation that remains, but the foundation cannot be expanded.

- This means that there's a limited time window to grow foundation skills in the PFC; the window opens at the onset of adolescence and closes when pruning is done.
- To beat the relentless pruning process, a teen needs to work on thinking skills early and often.
- Because most teens don't know about the PFC's thinking functions and the development process is unseen, silent, and gradual, many of them lack the motivation to do the work.

Recommended actions:

- For the parent: encourage your child to read Chapters 1 and 2, and then discuss the opportunities afterward.
- For the teen: based on what you know now about the vital functions of the PFC and how the development process works, reflect on how motivated you are to work on growing a smarter brain.

BONUS: Go to https://DrDennyCoates.com/bonus to download a free guide: "Maximum Teen Brainpower: 15 Essential Facts Every Teen—And The Adults Who Guide Them—Need to Know about Growing a Smarter Brain."

3

Your Child Is the Hero in His or Her Own Story

"Whatever we learn to do, we learn by actually doing it: men come to be builders, for instance, by building, and harp players by playing the harp. In the same way, by doing just acts, we come to be just; by doing self-controlled acts, we come to be self-controlled; and by doing brave acts, we come to be brave."

– Aristotle –
Greek philosopher (384 -322 B.C.)

While studying literature at the Duke University graduate school, I learned that stories have a universal structure, which can unfold in surprising ways.

The structure goes like this: the main character (hero) needs something, but unexpected challenges threaten to defeat him. If he's fortunate, a wise "guide" appears to show how to prevail. The guide encourages decisive action that will help the hero avoid failure and achieve success. But sometimes the hero struggles without a guide, or he fails to heed the advice.

Nearly all stories, novels, and movies use this structure to create the kind of drama that holds our interest.

For example, an alcoholic divorced dad gets a carpentry job that could allow him to continue making payments on his house. But his toolbox is stolen out of his truck while he's in a bar. This creates dramatic interest: will he recover his toolbox in time to begin the job?

Or as a couple prepares to celebrate their 40th anniversary, the husband gets an unexpected letter informing him that the body of his former sweetheart has been found on a mountainside, preserved under a sheet of ice—a lost love his current wife knew nothing about. What did this other woman mean to him? How will the wife handle the revelation, and will it change their relationship?

Always, the question is:

How will the hero deal with his or her challenges?

Will the deadbeat dad recover his tools and win the respect of his 10-year-old son? Will the wife come to terms with her husband's reaction to news about the woman who may have been the love of his life?

As the characters try to deal with adversity, we learn more about them. When the story arrives at a crucial turning point, we find out whether they have the inner strength to do what has to be done. If they do, they're rewarded and perhaps even changed for the better.

This structure also describes how events in our own lives unfold. Everyone lives their own stories—typically many stories—in real life.

I know this first-hand. In my life I had a successful career as an officer in the U.S. Army, retiring as a lieutenant colonel. I earned a Ph.D. from Duke University. I managed a creative, profitable business for over 30 years. Along the way, I designed several innovative online personal development programs that have been used by millions of people worldwide. I've written hundreds of articles and several books.

But while writing this book, I reflected on my adolescent years and wondered: Why did I become successful? What challenges did I overcome? Did I have guides in my life?

I was raised in the mid-twentieth century, when not much was known about parenting. Mostly, people raised their kids the same way their own parents had. My parents had average intellects. Mom was raised in a strict Mormon family, rebelled, and ran away from home before she could graduate from high school. Dad joined the Army during World War II after completing a year at a local college. They were both decent, hardworking people who raised eight children with a lot of love.

But nurturing thinking, sharing wisdom, and teaching life skills weren't a feature of our upbringing. I'm pretty sure my mom and dad never spoke the word "parenting" while I was growing up. They knew it was their responsibility to "raise a family," to provide food, shelter, and medical care, but they weren't able to save for our college education. When we finished high school, we were expected to leave home and make our way in the world.

Dad served two tours in Korea during my early childhood, meaning the job of raising my siblings fell mostly to my mother. By necessity she had to focus her attention on the younger ones. As the oldest child, I relished this benign neglect. When I was eight I'd hop on my bike and ride all over town—without telling my mother where I was going. It was a different time, but this freedom taught me to be self-reliant and self-confident. It helped shape the man I ultimately became—an independent thinker.

This kind of unconscious parenting could have opened the door to trouble. Aside from my parents' love and support, mostly my siblings and I raised ourselves, and the results varied widely. Even though all eight of us had the same two hardworking parents, each child turned out differently. Two of us graduated from college, but two others failed to graduate from high school. My next younger brother frequently got in trouble for things like vandalism, arson, and theft. After graduating from high school he enlisted in the Air Force but was later discharged for drug use. This troublesome personal history damaged his self-esteem, and he became a compulsive liar and ruthless con artist who dealt drugs. Ultimately, he was murdered in front of a convenience store in Miami.

I could have followed a similar path. Several years ago at a reunion with some high school buddies, we joked about how each of us had changed. At one point, one of my friends laughed and said to me. "Coates, you were always up to something."

I could tell by his tone that he wasn't talking about science projects. He was talking about mischief. This surprised me, because I thought I had been an outstanding kid, an Eagle Scout and straight-A student who worked to earn money and turned out well as an adult. Still, I couldn't put my friend's comment out of my mind, and a few years after the reunion I began recalling some long-forgotten memories of my early adolescence.

Like any normal 12-year-old boy, I began paying attention to girls, and my buddies and I would have hilarious conversations about sex. I remember being addicted to pinball machines. If I had an extra nickel in my pocket, I'd put it in the slot and try to win a free game.

I also perpetrated more than a few ill-advised pranks. On one winter day, my buddy and I positioned ourselves on a rooftop and launched water balloons at passing cars. When we finally scored a hit, we took off. Later, my scoutmaster surprised us by asking what we knew about the incident. He said it was a sub-freezing day, and a water balloon had cracked a windshield. We denied knowing anything about it. I remember shoplifting some comic books. And in the spirit of fun, my friends and I did other things that I'm not proud of. I was a typical teenager, full of angst and enthusiasm, impulsive, and not afraid of taking risks; and I did a lot of things without considering the consequences. To be honest, all this mischief could easily have led to hanging out with the wrong friends and getting involved in more serious offenses. In other words, I could have turned out like my younger brother.

But I got lucky.

For one thing, unlike my younger brother, I never got caught. For another, I was removed from this early environment.

In the summer of 1960 I was 15. My father, a chief warrant officer in the Army, had been reassigned to Germany, and our family would follow six months later when housing would be available. Meanwhile, I, my mom and my brothers and sisters were living in my dad's parents' house in Topeka, Kansas. I sometimes reflect on the love my grandmother brought to the horror of coping with all these kids in a small house. I had left all my boyhood friends behind in the small town of Waynesville, Missouri, and I was enrolled in Topeka High School for the fall semester, where my father had gone to high school when he was my age.

That summer my grandfather took a special interest in me, maybe because I was the oldest child and he knew I was a serious student. He was a typesetter at the *Topeka Daily Capitol*, which meant that his workday started at 6 PM and ended at 3 AM. Each morning he returned home to a quiet house and slept until mid-afternoon. In the afternoon he would come downstairs wearing a bow tie and a jacket. He was a small, slender man who loved to joke with his grandkids.

One afternoon Grandpa invited me to walk with him to work. He didn't own a car, so this walk was part of his daily ritual. Three blocks from the house we stopped at a corner store, where he bought a cigar. He introduced me to the man behind the counter. "Jerry, this is my grandson, Dennis. His family is staying with us until they can join my son Willy, who's been transferred to Germany. Do you remember Willy?"

"Yes, but that was quite a while ago."

After we left the store, Grandpa pointed to a brand-new red-and-white Buick parked at the curb. "What do you think of that car?" he asked.

I didn't know how to answer. Was he testing my knowledge of cars? "I like the color," I said. "I bet it's expensive."

"Dennis, somebody designed that car. Or rather, a group of people. Think about it. They learned how to design cars, got hired by the Buick company, and look what they accomplished."

I had never thought of cars that way, that real people had become smart enough to make the cars I saw every day.

"You could do that someday," said Grandpa. "If you wanted to."

I was pretty sure I didn't want to, but the idea that something like that was possible was a new thought.

Grandpa lighted his cigar. "I don't inhale," he said. "Jerry's an interesting fellow. He's owned that store for thirty or forty years. I see him every day. He's just as smart as the guys who design Buicks, but he's smart about other things. There's a lot to know about owning a store. You could operate a store of your own someday if you wanted to."

I wondered about that. Where do you get all the stuff that's sold there? How much do stores cost? It seemed beyond my experience. "I don't know, Grandpa."

As we continued our walk down 10th Street, the state capitol building came into view. "See that guy mowing the grass? His boss has a contract to keep the grounds looking great. It's a hard job. How would you like to spend all day every day cutting grass?"

As we walked by a bus stop, a bus pulled up. Grandpa waved at the driver. "I sometimes take this bus. The driver there is Howard Sears. He's a family man, two boys and a girl. I think one of the boys is about your age. Maybe you'll meet him in school. I don't know how many times Howard drives this route every day. It's a hard job, but I'm glad he does it. And I'm glad there's somebody willing to drive the city busses and mow the Capitol lawn."

When we arrived downtown we entered a drug store. "You want a soda?" Grandpa spotted another friend. He was also dressed in a suit and tie. They shook hands. "Jim, good to see you," the man said. "Sit down. Who's this young man with you?"

"Alf, Dennis is my oldest grandson." Then he turned to me. "Dennis, Alf used to be the governor of Kansas. He ran for President 25 years ago."

Twenty-five years seemed like ancient history to a 15-year-old boy. Why would someone like him be in a drug store in Topeka? How did he come to be friends with my grandfather? It was a lot to comprehend.

When we got to the newspaper building, Grandpa talked about Oscar Stauffer, the owner. "His company owns the *Journal*; they bought the *Capital* a few years ago. He's a very accomplished businessman. But it's time for me to get to work now. You know the way home. We'll talk more later."

More than half a century later, I understand that the summer of 1960—and especially this walk—was a turning point in my life.

In a real sense, *my grandfather was my guide*. He made me realize that while some people prepare themselves to do great things, others don't. He helped me believe that I could do something significant, but it was up to me what that would be.

How lucky was that?

When our family moved to Germany, I found myself in a strange new world and in a different high school, where I knew no one. While I had been both an A-student and prankster before, I had to start over; and the chance that I might unwittingly walk down the wrong path was eliminated.

Amazing luck.

The changes put me in a serious frame of mind, and I began thinking about my future. I focused on my studies and became more responsible. As the oldest child, I was the go-to caretaker for my siblings. I attended church twice on Sunday with my family, and I even taught a Sunday school class. I babysat, I caddied, and I saved my money. I was the only kid in my high school class who made straight A's, term after term. I was a wrestler and captain of the golf team. I was vice-president of the student council and valedictorian of my graduating class.

These credentials were enough for the governor of Kansas to offer me an appointment to the U.S. Military Academy, *even though I had never been a permanent resident of Kansas.*

Another amazing stroke of luck.

I had wanted to be a West Point cadet for several years. When I was in middle school in 1958, I happened to read a magazine article about a West Point cadet named Peter Dawkins.

Dawkins was "first captain," the highest cadet rank in his class. Academically, he was a "star man," eventually graduating 8th in a class of over 500. And he won the Heisman Trophy as a running back on the Army football team. I was impressed that anyone could achieve so much. I wanted to be like him. Coming across this article was another key moment in my life. In a way, the article served as another guide.

More good luck.

At West Point, attending class was mandatory—over 20 credit hours per semester. I found out that a lot of guys my age were smarter than I, a humbling discovery. The feeling that I might be less than excellent motivated me to study as hard as I could, in spite of the pressures of cadet life, which were many.

My courses were challenging. Even though after graduation I used very little of what I learned, the instructors made me think. I didn't know it at the time, but by trying like crazy to get good grades, I was giving my PFC an amazing workout. Courses like advanced mathematics, civil engineering, fluid mechanics, strength of materials, quantum physics, and nuclear engineering—most of them required—involved a lot of logical analysis and problem solving.

I took enough math courses to earn a major, though back then the Academy awarded only general engineering degrees. The math courses required us to study a procedure at night, and the next day each cadet had to solve three problems on a blackboard without referring to the text. At random, one of us would be called on to explain his work in front of the class—for a grade. Military history was all about analyzing why battles were won and lost. Even English classes guided me not just to enjoy literature, but to understand the underlying themes.

From a brain development standpoint, four years of cadet life was one of the best things that ever happened to me. I wasn't conscious that it was making my brain smarter, but during my teen years and early twenties, I wired my PFC for a myriad of executive functioning and critical thinking skills, all of which had a profound impact on the rest of my life.

So that's the story of my adolescent journey. Yes, when given the opportunity, I worked hard. But much of my good fortune happened because of factors beyond my control. My grandfather and the example of an illustrious West Point cadet inspired me.

I had some amazing good luck.

In the life of every adolescent child, a similar story is unfolding. It involves a journey, challenges, drama, and enormous consequences.

The incident that triggers a young adolescent's story is puberty, the beginning of a final, 12-year period of physical and mental growth. The challenges every teen faces are formidable. Will they become the kind of person they want to be? Will they stand up to peer pressure? In a world they poorly understand, will they stay out of trouble? Will they take their studies seriously? Will they adopt a work ethic and learn life skills? Will the child grow up to be an individual who is strong in character and mind who achieves happiness and success in life? Or will the struggling teen grow up to be a struggling adult? Will a guide—

a parent, relative, teacher, or mentor—intervene to show the way?

Or will the child get lucky? Will he or she decide to pursue activities and learning that stimulate the development of a robust intellect?

As in every phase of child development, to establish vital thinking skills a child has to use them repeatedly. Every young person has about a dozen years to create the neural pathways in the PFC for foundation thinking skills that are vital for success in life and work.

The turning point of every teenager's story is: *Will they be the beneficiary of helpful influences? And will they do the work?*

Before the 21st century and throughout human history, no teenager was told about their developing PFC. No parent or child was aware of its functions or that this area of the brain was ready for basic development during adolescence. Therefore, young people couldn't consciously build the kind of thinking skills that make a huge difference in life. If teenagers somehow got involved in activities that established thinking skills, it wasn't because they wanted to wire their PFC.

While I'm grateful for the good luck I had while growing up, the point of this book is that:

The development of an adolescent's brain no longer has to be a matter of luck.

In the past, not all kids got lucky; all too many of them didn't grow up to be adults with superior minds.

Kids born in the 21st century are literally the first generation in human history to have the opportunity to understand what's going on in their growing brains and consciously wire their PFC for foundation thinking skills. Today any teen can proactively achieve the best possible outcome. If they are made aware of the insights in this book, they can, with help from their parents and other caring adults, consciously and intentionally make the best-case scenario happen. *They can literally take luck out of the equation.*

Also, the parents who hope for the best for their teen are heroes in their own story: *They can mentor the child successfully so that he or she leaves home with robust mental capacity.* The next six chapters explain how.

Major takeaways from Chapter 3:

- Every teenager (and every parent) is pursuing a journey that unfolds according to classic story structure.
- Whether the young person prevails against the formidable challenges of adolescence depends partly on whether a guide intervenes to show the way.
- The kind of adult a young person becomes has a lot to do with how much time they spend during youth exercising thinking skills.
- Some teens spend more time exercising their PFC than others and become adults who have more intellectual capacity than others.
- Until recently, kids had to make the right choices without knowing that what they were doing was linked to their brain development.

- If an adolescent child applies the recommendations in this book, he or she can consciously grow a smarter brain.

Recommended actions:

- For parents: Read the remaining chapters to find out what you can do to encourage activities that will help your child develop maximum brainpower.
- For the teen: Read the remaining chapters to find out what you can do to develop your PFC.

Part Two

WHAT PARENTS CAN DO

What does it mean to be smart?

It could mean that you've stored a lot of information in memory.

Or it could mean that you've acquired the kind of thinking skills that enable you to use this information to solve problems, make decisions, and get things done.

The wonderful thing about being a teenager is that a kid's brain is ready to be wired for all these basic thinking skills, and teens can consciously choose to do the kind of work that will make it happen.

While Part One was about how basic brain development works, Part Two is about *action*—three things adults can do to help a child develop a superior mind.

Chapter 4 asks adults to let the young person in on the secret—sharing the insights of this book so they will know how growing a smart brain works and how to make it happen.

Chapter 5 explains how to ask the kind of questions that will stimulate a teenager to think—to exercise the PFC.

Chapter 6 describes a new, 21st century problem: how to protect a young brain from excessive screen time, which can disrupt a teen's efforts to grow a smart brain.

4

Let Your Child in on the Secret

"Knowledge is power."

– Francis Bacon –
British philosopher (1561-1626)

A recurring theme throughout this book: no adult can wire a teen's PFC for them. *The young person has to do the work.* And the motivation to accomplish this has to come from within.

But how can you expect a child to commit to a journey of exercising important thinking skills if he or she has no idea what's going on?

An essential first step to nurturing this motivation:

Let the kid in on the secret.

As scientist and philosopher Francis Bacon said, "Knowledge is power." I believe that if adolescents understand how high the stakes are, what the possible consequences are, and what they need to do to develop the kind of superior mind that will make them successful, many of them will want to make better choices. I'm not aware of a single downside to a teenager's having this kind of self-awareness.

Also, "the work" to build the PFC will be inherently motivating. It will involve academic courses and extracurricular activities that are interesting and fun—not special "brain-building" exercises.

Puberty launches a period of significant physical development, much of which is necessary for reproduction. While the changes are easy to notice, they don't manifest at the same age and in the same way with every child. But every young person needs to know what's happening to them.

What's not visible or obvious is the developmental process that's taking place in the adolescent brain. A lot rides on whether a young person learns about these changes. They can't make conscious choices that will benefit them if they don't understand what's happening. Giving adults and teens this information is the purpose of this book.

An important question: When are young people ready to receive this empowering information about their growing brain?

In many ways, middle school is a whole new ballgame. With puberty underway, young people aren't little kids anymore. They aren't adults yet, either, and they won't be for another ten years or so. As they begin the physical and mental journey towards adulthood, they have a lot to learn. Is a twelve-year-old ready to think about growing a smarter brain?

My bias: the sooner the better.

The speed with which pruning advances isn't well understood. Very likely, the elimination of unused brain cells and connections in the PFC varies from child to child. It's possible that for some teens, most of this pruning will have been accomplished before the end of high school. Most likely, the kids who start working on executive and critical thinking in middle school will get a head start on social and academic success.

Also, a powerful force will grip their attention and motivation in middle school. I'm talking about the youth culture and the all-consuming desire to be liked and included. This is a "right here, right now" need, which has very little to do with preparing for their future as adults.

Focusing on the future—even a month into the future—can be hard for a preteen. It's hard for a sixth-grader to imagine what seventh grade will be like or to imagine how hard it will be for them to get ready for next levels such as high school, college, or life as an adult.

Will they use good judgment and make the right choices? Will they resist peer pressure and say no to alcohol, drugs, sex and other temptations? The sooner they get involved in activities that will improve their thinking skills, the better.

It can be hard for a parent to explain the important aspects of adolescent brain development. The best approach for a concerned parent: *share this book with your child.*

After he or she finishes a chapter, you can talk about it together—what thinking skills are, why they're important, and what your child can do to build the strongest possible intellectual foundation.

Once a young person appreciates that they are on a path towards becoming an adult, and that their school years are meant to prepare them for this, they'll become ready to consider insights about their brain development.

In a way, giving young people information about their developing brain is like having talks about sex. They need to know what's happening to them so they can make beneficial choices. Using this book as a guide, you don't need to worry that you'll leave out something important. It translates brain science research into a nontechnical explanation that anyone—even a preteen—can understand.

Major takeaways from Chapter 4:

- A child who isn't aware of adolescent brain development won't be motivated to work on thinking skills.
- If you want your child to choose to grow a smarter brain, you'll need to share what you've learned from this book.
- Middle school students are smart enough to understand this information.
- However, they'll need a willingness to think about their future to be receptive.

Recommended actions for parents:

- Engage your child in talks that help make the future seem real to them.
- When you've finished reading this book, encourage your child to read it, too.
- Have them read a chapter or two, then discuss it together. Instead of summarizing main points, ask what your child thinks.

5

Encourage Your Teen to Think

"Children must be taught how to think, not what to think."

– Margaret Mead –
American anthropologist (1901-1978)

Imagine that your child is now an adult, working at a new job and building a life. You know that most of what matters is going to be hard:

- Finding fulfilling work
- Dealing with managers and coworkers
- Building lasting relationships
- Managing finances
- Owning and maintaining a home
- Raising a family
- Learning new life skills

- Solving the problems that come up every day

Some young adults struggle with relatively easy tasks, such as fixing dinner or doing laundry. So when your child runs up against the inevitable frustrations, which would you prefer that they do:

(a) Call and ask for help
(b) Handle it without your help

If you're missing your child, you might be delighted to have the call, whether the news is good or bad. You have a treasure trove of life experience, and you may feel that sharing your wisdom or advice is an expression of parental love. If so, you might be tempted to choose (a).

But the better answer is (b). Assuming your own parents are still alive, do you lean on them for solutions whenever you have a crisis? Very likely you just confront the problem, think it through, decide what you want to do, and then do it. And if it doesn't turn out the way you planned, you learn from your effort and try something else. Because that's what adults do.

The best-case scenario is that your child becomes an independent adult, confident that they can deal with whatever comes up next.

But this capacity doesn't suddenly turn on just because they're fully grown and gone. The ability to create a life in today's world will require character strength, life skills, and a variety of thinking skills, such as those described in Chapter 1. The only way to acquire these skills is to exercise them repeatedly during the years before they leave home.

The best way to help your child isn't to shield her from frustrations. She needs to get used to resolving issues on her own. You can be there for her with empathy, support, encouragement, and by getting her to think for herself. In short, recognize that your child is an adult-in-the-making, and you have a limited amount of time to help her prepare.

The basic approach is simple:

Ask open-ended questions to get your child to think

An open-ended question is one that can't be answered with a simple "yes" or "no" or a word or two. This fairly simple communication technique is the opposite of using your superior knowledge, skills and experience to give answers and instructions. While doing the thinking for them may resolve the immediate issue, it won't help your teen create the brain circuits for thinking for herself.

An open-ended question can take many forms, depending on the situation. Here are ten typical opportunities for helping your child wire her brain for patterns of good judgment and decision-making, along with examples of open-ended questions.

When your child doesn't UNDERSTAND HOW OR WHY...

Things happen for a reason. Young people may be aware of what's going on around them, but they may not ask themselves why. If they aren't coached to be curious during adolescence, this habit of mind may never get wired, and they may not be inquisitive as adults.

A parent can encourage a child to ponder how things relate to each other or to understand the connection between cause and effect. The key is to get your child to notice ordinary things and ask "why."

For example, say you and your child are getting in the car to run an errand. If he complains about having to wear seatbelts, you could ask: "Why do you think the state made it a law?" Or if you notice that people are getting their gas tanks filled at a station when gas is cheaper only two blocks down the street, you could point this out and ask: "Why do you think they're doing that?

Some generic questions...

- "What do these things have in common?"
- "In what way is [A] different from [B]?"
- "Why do you think [A] is more interesting than [B]?"
- "How important is this to you?"
- "Why do you think this situation exists?"
- "What does this mean?"
- "How does this work?"
- "Why do you think this happened?"
- "Why do you think she said that?"
- "What was the artist trying to express?"
- "Why do you think I want you to do this?"

The idea is to encourage your child to think about why the world works the way it does.

At breakfast, a mother is reading the lifestyle section of the newspaper. She asks her daughter, “Have you noticed how fashions change every year? If a style is so appealing, why do you suppose designers always change them?”

“To get people to keep on spending money on clothes?”

“That makes sense. How do you feel about that?”

Dad and his son are playing catch with a football. During a break, Dad asks, “Hey, can you feel how your cheek that faces the sun is warm, and the other side feels cooler?”

“Yeah.”

“It’s pretty hot out here. Why do you think one cheek feels cooler?”

“I guess because the sun is hot and one cheek is facing the sun and the other isn’t.”

“Right. But the sun is a long way from here. How is it that the sun is so far away but can still make your skin feel warm?”

You may be asking your child to do a kind of thinking he’s not used to. So don’t be surprised if you sometimes get the answer, “I don’t know.” This is an honest answer, but a lazy one. The key is to not give your own explanation. Instead, encourage your child to think it through or search for an answer. Say something like, “Yeah, it’s not so obvious. But I want to know what you think.”

When you want your child to FOCUS ON THE FUTURE...

Most of the time, kids are focused on the present. They don't often ponder the future.

You want them to learn the kind of life skills they'll need as adults. But they won't be motivated to do that if they can't imagine a future when these skills will become necessary. To take schoolwork seriously, they have to understand how it might help them later when they're on their own. To avoid a dreadful consequence, they need to imagine that it could happen in the future. Your child might want to save or invest their money, but only if they've thought about the future reward for doing so. To set a goal, a child has to have something worth working towards—something they can possess in the future. Many of your efforts to parent will stall if the future doesn't seem real to your child.

To get your child used to imagining the future, try asking questions like these:

- "What are you planning to do this weekend?"
- "What do you want to do when the school year is over?"
- "What plans do you have for the money you're earning?"
- "How do you see yourself using the car once you get a license?"
- "What will you do after you graduate from high school?"

When you want your child to FORESEE CONSEQUENCES...

Knowing that doing something causes a certain result requires imagining something that hasn't happened yet. But if a child has observed cause and effect in the past, he or she can apply the memory of this to a similar situation in the present.

For example, say your child wants to take a summer course at the community college instead of spending two weeks visiting relatives. "How do you think the course will help you?" "How do you think your relatives will feel if we show up without you?" Or maybe some friends want your child to go white-water rafting with them. "What do you think this kind of adventure will be like?" "What skills will you need to do it safely?"

The idea is to get your child to think about what causes things to happen and to imagine what might happen as a result of their actions. This mental skill is vital to impulse control, good judgment and decision making.

Typical questions...

- "In this situation, what do you think will happen?"
- "What do you hope to get out of this?"
- "How will this help you achieve your goal?"
- "If you do this, what could happen next?"
- "Can you think of another way to get what you want?"
- "What could get in your way?"
- "Why doesn't this feel right to you?"

- "Based on what you know about her, what do you think she'll say?"
- "What bad things might happen if you do what you're thinking about doing?"

The key is to avoid doing the thinking for her, which is what you may have had to do when she was younger. But a teen is preparing to work through tough problems and needs to exercise the parts of her brain involved in foreseeing future situations.

When your child needs to SET GOALS AND MAKE PLANS...

Your child may decide to try out for the track team next year. Or she might want to earn enough money to buy the latest smartphone. To get there from here will involve a number of steps. If she has a plan and follows it, she's not only more likely to succeed, but her self-esteem will grow if she does.

Once a child is comfortable thinking about future goals, she can envision what she has to do to make it happen.

Typical questions...

- "What would you like to achieve this year?"
- "What will you need to do to get that?"
- "What's most important to you?"
- "What do you want to do after graduation?"
- "What are your other options?"
- "How much will this cost you?"

- "What are the advantages and disadvantages of your plan?"
- "How will doing this benefit you?"
- "What will you need to do to get that?"
- "What would be your best first step?"
- "What's the hardest challenge you'll have to face?"
- "If this happens, how will you deal with it?"
- "What help or resources will you need?"
- "If you go with this plan, how do you think it will work out?"

It can help if your child writes down the action steps. If she writes each action on a post-it note, later she can easily rearrange them in time or cause-and-effect sequence. Listen to clarify what your child says, while avoiding lecturing, explaining, or trying to solve the problem. In other words, ask questions to get your child to do the planning.

One good way to teach planning skills is to involve your child in planning family events.

Mom – "Are you looking forward to our trip to the mountains?"

Son – "Definitely."

Mom – "I'm going to need your help."

Son – "What do you mean?"

Mom – "You know how busy I am at work right now. I'd like you to figure out what we're going to need for the trip—from start to finish. We leave a week from Friday. I need you to pull it all together so we're ready to go. Can you do that?"

Son – “Sure.”

A few days later, your child asks, “Do you think I should fill the gas tank?”

Mom – “What do you think?”

Son – “I guess so. And what about food? What do you want me to pack?”

Mom – “Well there’s four of us and we don’t come back until Monday night. Figure it out and make sure we have enough.”

When you want your child to EVALUATE something...

One aspect of good judgment is the ability to appreciate the better option—the ability to assess the value of something. You can help your child become a more discerning thinker by asking questions like:

- “How are these three options different?”
- “What does this do for you?”
- “Why is this better than that?”
- “What are the advantages of doing this?”
- “What are the downsides?”
- “Why do you like that?”
- “What do you admire about this?”
- “What are its best features?”
- “What are its flaws?”
- “How well does it do what it’s supposed to do?”
- “How much is this worth?”
- “How safe is this?”

- "How is this good for you?"
- "What are the strengths of that argument?"
- "Why is that movie one of your favorites?"
- "What message do you think the artist is expressing?"
- "How well does that person do this?"
- "What do you think about how well this is made?"

When you want your child to BE MORE SELF-AWARE...

As young people begin to seek independence, they want to fit in with their peer groups. They'll be concerned about how others see them. As they seek an identity, they might experiment with roles and looks, or they might begin the serious work of creating who they want to be. Parents can help by asking questions that encourage self-awareness:

- "What's your opinion about that?"
- "When this happens, how do you feel?"
- "What makes you feel happy?"
- "What makes you angry?"
- "What are you good at?"
- "How do you feel about yourself?"
- "How do you think others see you?"
- "What about yourself would you like to improve?"
- "What is your No. 1 goal right now?"
- "What's most important to you?"
- "What do you look forward to most these days?"
- "What would you like to be doing five years from now?"

When you want your child to STAY FOCUSED...

Because the PFC is heavily interconnected with other parts of the brain, it receives real-time input and can direct the activities of other areas of the brain. For example, when we dream, remember or imagine, the PFC sends signals to the visual cortex to create images. If the amygdala sends fight-or-flight alarm signals, the PFC can quickly evaluate sensory input and decide whether to send a signal back to calm the amygdala. If the senses notice something, the PFC can evaluate how important the information is, and then direct the brain to focus on it or pay attention to something else. In this way, it's possible to learn how to manage attention, ignore distractions, and stay focused. These questions can help a child wire the PFC for the skills of staying focused:

- "What you're doing right now—how important is to you?"
- "What's your top priority for today?"
- "During the next hour, what do you want to accomplish?"
- "What can you do next to get back on track?"
- "What should you be paying attention to right now?"
- "What's the payoff for finishing this?"
- "What can you postpone until later?"
- "When somebody wants you to do something, what's a nice way to say no?"

Guide your child to THINK THROUGH PROBLEMS AND CHALLENGES...

Say your son is upset because he forgot that he promised to help a friend work on a project, and then arranged to meet someone else at the same time. Or he wants to buy an expensive item and doesn't have the money to pay for it. Or he put off writing an essay and now it's due the next day.

Typical questions...

- "What's the real problem?"
- "What do you think went wrong?"
- "How have you dealt with situations like this in the past?"
- "What other options do you have?"
- "What's possible in this situation?"
- "Which is more important to you, [...........] or [...........]?"
- "Is there an even better way to [...........]?
- "Why do you feel that this option is the best one?"
- "If you do that, how will it meet your needs?"
- "How will doing this benefit the other people involved?"
- "What are the potential dangers?"
- "What could this end up costing you?"
- "Is there an option that will be acceptable to everyone?"
- "What does your gut tell you?"

The idea is to ask questions that lead your child to go through the typical steps of problem solving: identifying the real problem, thinking of possible solutions, considering the pros and cons of each option, and picking the best one. Even though you want the best for your child and have lots of experience, keep in mind that it's your child's problem, not yours. Recognize the opportunity for your child to exercise problem solving skills, and resist the temptation to push your idea of the best solution.

Mom – "You seem down today."

Daughter – "I didn't get the lab assistant job."

Mom – "I'm sorry, dear. I know you prepared for it."

Daughter – "They picked my friend Ella."

Mom – "How do you feel about that?"

Daughter – "Well it's nice for her. But I was counting on the money. I was going to use it to pay for the Paris trip. Now I don't know what to do."

Mom – "I know you had your heart set on it. What other options do you have to get the money?"

Daughter – "I don't know. None."

Mom – "What can you do to find out?"

Daughter – "I guess I need to check the want ads."

Mom – "Today's paper is still on the dining room table."

Daughter – "Or I could just go to the mall. Maybe some of the stores still have 'We're Hiring' signs in the window."

Mom – "Does that sound promising?"

Daughter – "I think I'll go down there. I remember seeing signs at some of the clothing stores. I might get lucky."

Mom – "Sounds like it's worth a try. Do you have other options?"

Daughter – "Well, the lab assistant job is gone. But maybe the school has some other jobs open. I could check."

Mom – "So you're going to check both?"

Daughter – "Yeah. And while I'm at it, I'll talk to Mrs. Bath, the guidance counselor. She's kind of plugged in."

Mom – "Go for it, Kiddo."

When you want your child to LEARN FROM EXPERIENCE...

It's often said that experience is the best teacher and that the value of mistakes is that you can learn from them.

There's a lot of wisdom in this. Kids sometimes experience success; but do they understand what worked, so they can repeat the success in the future? And what about mistakes? Kids don't always learn from what happens to them. A young person could easily go from situation to situation without giving it much thought and without learning a thing.

When your child makes a mess of things, instead of getting mad, criticizing, or lecturing, ask these open-ended questions to help him analyze what happened:

- **What happened?** *Who did what? What was the sequence of events?* The details of an event need to be recalled in order to make sense of them.
- **Why did it happen that way?** *Cause and effect? Your motives? What helped or hindered?* Things don't just happen. They happen for a reason. To imagine a better way to handle a situation like this, your child needs to understand why things happened the way they did.
- **What were the consequences?** *Outcomes? Benefits? Costs? Problems? Resolutions?* Your child needs to appreciate the impact of what happened in order to want to handle it differently in the future.
- **How would you handle a similar situation in the future?** *What lessons did you learn? What basic principles?*
- **What's your plan?** *How are you going to apply this lesson in your life?*

The key is to get good at spotting a "learning moment"—when something significant has happened to your child. When you sense excitement, agitation, anger or depression, this is your cue to take a few moments to ask the questions. The exchange should take the form of a discussion, not a lecture. Let your child do most of the talking. The key is to encourage your child to give the answers, even if the insight is perfectly clear to you and you want to make sure your child gets the lesson.

When you ask a question, listen to understand what your child is telling you.

Dad – "What's wrong?"

Son – "Nothing."

Dad – “You slammed the front door just now.”

Son – “So I’m sorry already!”

Dad – “You seem upset about something.”

Son – “No I’m not.”

Dad – “Tell me what happened.”

Son – “It’s no big deal. Just Jerry being a jerk.”

Dad – “What did he do?”

Son – “I lent him my book on fly fishing and when I asked him to return it he said he already gave it back. But he didn’t.”

Dad – “So why do you think he said he gave it back?”

Son – “I don’t know. Maybe he just wants to keep it. He’s had it for almost a year and never returned it. Now I want it back.”

Dad – “That’s a long time.”

Son – “Yeah.”

Dad – “How can you get him to return it?”

Son – “I don’t see how. He says he doesn’t have it anymore. Maybe he forgot where it came from and gave it to somebody else. I don’t know.”

Dad – “Frustrating.”

Son – “It makes me mad.”

Dad – “I can imagine.”

Son – “It was my favorite fishing book. Grandpa gave it to me. I thought I could trust Jerry with it.”

Dad – “I’m sorry it didn’t work out. How can you keep something like this from happening again?”

Son – “I’m done letting Jerry borrow my books. He can get his own books from now on.”

Dad – “Is this how you want to handle it?”

Son – “Maybe I shouldn’t be lending any of my books.”

Dad – “They’re important to you, aren’t they?”

Son – “This one was special. Some of the others I don’t care about.”

Dad – “So you’re saying maybe you won’t give out the books that matter.”

Son – “Maybe.”

Dad – “Do you think this will work out for you?”

Son – “People might think I’m stingy. But they’re my books. I don’t have to let people borrow them.”

Dad – “Okay.”

Son – “And maybe I’ll get another copy of the fly fishing book.”

Dad – “Sounds like a great idea.”

When you want your child to SELF-REGULATE AND CONTROL IMPULSES...

When a kid gets angry, he might express rage by striking out or saying something hurtful. While this is a common behavior of small children, it can be disastrous for adults to act this way. It's possible for young people to learn to be aware of their emotions, consciously pause until the emotions subside, and consider more effective responses. These mental activities take place in the PFC. These questions can help your child learn to stay cool in heated situations:

- "What are you feeling right now?"
- "Why do you feel this way?"
- "What do you feel like doing?"
- "If you do that, what could happen?"
- "What would happen if you wait to say or do something?"
- "What can you do to calm yourself?"
- "Why did this person do this?"
- "What can you say or do that will benefit you the most?"

All the questions suggested in this chapter are variations of *What do you think?* When there's an opportunity to ponder, discover, think ahead, learn, plan, or deal with problems, open-ended questions will get your child to think for himself. I've said many times that your child has to "do the work" to wire his or her own PFC and that this will require many repetitions of thinking skills. Asking open-ended questions is a simple but powerful way to help make this happen. And each time you ask for your child's thoughts, you'll send the message that you value them, which will build your child's self-esteem.

Major takeaways from Chapter 5:

- The best way to get a young person to think is to ask an open-ended question.
- Asking a variation of "What do you think?" is far more fruitful than giving advice, instructing, or handling a situation yourself.
- There are specific questions that relate to common learning situations.

Recommended actions for parents:

- Adopt an attitude of genuine curiosity about your child's thought processes.
- Bookmark this chapter as a reference for typical questions.
- After asking a question, follow up with listening rather than giving a better answer.
- Make a concentrated effort to ask open-ended questions until it becomes a habit.

6

Avoid Excessive Screen Time

"We know that actual brain damage occurs from excessive Internet and video game use that looks remarkably similar to that from drug and alcohol abuse."

– Victoria L. Dunckley, MD –
Integrative psychiatrist, author of *Reset Your Child's Brain*

There are two elements in the teen culture that have the potential to interfere with a developing PFC. One is substance abuse—alcohol, drugs, and nicotine. These foreign substances can disrupt normal PFC development. I address this threat in Chapter 8, written for teens, because only the teens themselves can make a commitment to abstain from these substances while they're young.

The other peril is excessive screen time, which also can disrupt normal adolescent brain development. In this case, parents have given the devices to their kids, and so they have more control over how their kids use them.

While this chapter is focused on how screens can hinder development of the PFC, screens can be harmful in other ways. Readers who want to learn more about screen addiction, distraction, self-esteem, bullying, social skills, pornography, and other dangers can read Thomas Kersting's *Disconnected: How to Reconnect Our Digitally Distracted Kids*.

Most parents don't understand the harmful impact of excessive screen time. The reason is that screen technologies are such a new aspect of our culture; and it's the wonders of these technologies, not the negative side effects, that are heavily promoted.

These amazing 21st century devices have advanced through our culture in what seems like the blink of an eye. As helpful as they are, they have caused unintended negative consequences. In this chapter I'll explain how the devices do their damage and what you can do to protect your child's brain. I suspect that for many people (especially the kids) this information is unexpected and unwelcomed. The hard, inconvenient truth is that *excessive exposure to screens can disrupt your teen's developing brain*. It can happen two ways: (1) hours of intense interaction with action-based or violent video games, or (2) repeated exposure over months or years to a variety of milder screen media.

Excessive screen time is a new, 21st century problem brought on by the recent, wondrous changes in information technology.

You may wonder how something so amazing and beneficial, welcomed by most of our society, can be dangerous to your child. But the harmful impact of accumulated exposure to electronic screens is real, severe, and can last a lifetime; and your child is too important to become a victim.

The rapid evolution of electronic media has taken us by surprise. In fact, your teenager belongs to the first generation of young people to be in the crosshairs of this danger. Today we find ourselves in a world vastly different from the one in which most parents grew up. For example, my boys were teenagers during the mid-80s. When they grew up, game consoles, smart phones, laptops, tablets, and e-readers didn't exist. The primitive video games they played involved goofy symbols moving around on a tiny black-and-white PC screen. The Internet wouldn't become a commercial service until they were in their 20s. Only a few schools had portable computers and most of those sat unused in the school library. Nobody used email. Nobody "texted." Nobody took "selfies" with a phone that fit in their pocket. You took pictures with a camera that had a roll of film in it that had to be processed at the drug store. You put the pictures you wanted to save in an album or a box. The prevailing social medium was the telephone, which connected you to other people by wire over telephone poles. If you wanted to "friend" someone, you called them or visited them in person. If you wanted to make a call away from home, you had to find a phone booth.

Today, most phone calls are wireless, relayed by distant radio towers or by Wi-Fi devices connected by fiber. What we used to call a telephone is now called a landline, and these are disappearing from homes. The computers people carry in their pockets are many times more powerful than the computers astronauts carried with them to the moon. They're smaller than a pack of cigarettes, and you can use them to make phone calls, send messages, get directions to distant destinations, do research, listen to music, watch sporting events or TV shows, buy and sell stocks, access more information than exists in the Library of Congress, and take a high-definition digital photo and instantly send it to anyone, anytime, anywhere in the world. They're as expensive as a desktop computer, and your teenager would like one of these for Christmas (if they don't already have one).

And this aspect of our culture is rapidly advancing in unexpected ways. There is now something called "esports." Dozens of small universities promote their own varsity esports programs, offering scholarships worth thousands of dollars to talented young gamers. There's even a governing body: the National Association of Collegiate Esports. One small university reported that esports is their only sports program. Many parents are even hiring gaming tutors to improve their kids' chances of winning, believing that losing while playing a popular video game can't be great for self-esteem, and maybe winning could lead to one of those scholarships.

The publishers of video games are sponsoring tournaments with big cash prizes. Some gamers make over $100,000 in prize money from these competitions. Virtual reality gaming competition is next. It's likely that university esports competitions will become features on ESPN, like poker.

In August, 2018, David Katz, a champion gamer, was eliminated in a Madden NFL gaming tournament in Jacksonville, FL. Frustrated by his defeat, he got a gun and started blasting away, killing two gamers and wounding 10 others. One can only imagine what was going on in his brain when he then turned the gun on himself. According to his mother, he was so addicted to gaming as a teen that he would refuse to go to school or bathe. "When I took his gaming equipment controllers away so he couldn't play at 3 or 4 in the morning, I'd get up and find that he was just walking around the house in circles."

A kid can't hope to achieve the level of competence to win an esports scholarship or tournament without being an addicted gamer. And addicted gamers have so much cortisol accumulation in their brains that brain cells are dying. Hyper-aroused teens are hard to control, and they neglect the normal, healthy aspects of growing up, such as sleep, hygiene, physical fitness, friendships, learning, time spent in nature, and competing in real sports.

The Warning Flags Have Been There

Experts have been raising issues about screen time for 25 years. It started with Jane M. Healy, author of *Endangered Minds* (1990), who summarized research showing that television was having a verifiably negative impact on the developing brains of children. At the time, parents saw TV as a blessing, using kiddie programs as a kind of babysitter. Healy pointed out that watching TV is a passive activity. Because it grabs kids' attention with visually exciting images and fast-paced, rapidly shifting content, it conditions them to prefer getting information this way rather than, for example, listening to an adult or reading a book.

Her later book, *Failure to Connect* (1998), challenged computer-based education, claiming that yes, the new learning technologies were indeed attention-grabbing. But no, these programs were doing little or nothing to improve children's ability to learn. And like television, these programs were training kids' brains to be bored with books and the kind of classroom learning that requires focus and persistence.

Nicholas Carr's best-seller, *The Shallows* (2011), explains that throughout history each new information medium, such as clay tablets, scrolls, printed books, radio, film, television—and now computers and the Internet—had to be used in a new way, thereby rewiring the human brains that engaged with it. Carr claims that while the Internet gives us faster access to a lot more information, it displays so many options that the information-gathering process is unfocused and distracting. It causes us to shift our attention more often and come away satisfied before we've explored a topic thoroughly. In other words, habitual use of the Internet degrades the habits of focus and concentration that are necessary for in-depth learning and problem solving.

In his newly updated classic, *Why Do They Act That Way?* (2014), child psychologist David Walsh describes the ambivalence and controversy surrounding teen use of electronic devices. On the one hand, social media can extend and enrich relationships that have been nurtured in real life. And the Internet can give your child access to knowledge and information far beyond anything you found in libraries when you were a kid. On the other hand, when young people multi-task with electronic devices while studying, they get less done and create the habit of needing entertaining distractions while learning. Also, they can become addicted to the rush of pleasure when a text message arrives or they earn the next level on a video game. It's not uncommon for a teen to spend over 10 hours a day in front of electronic screens, and the result can be sleep deprivation, poor academic performance, social isolation, and lack of exercise leading to obesity, along with health problems such as diabetes. Perhaps more concerning, Walsh, a long-time critic of violent video games, cites research proving that excessive exposure to violent gaming content can lead to callous and aggressive behavior. As he says, "Whatever the brain does a lot of is what the brain gets good at."

Neuroscientist Dr. Amy Nutt, author of *The Teenage Brain* (2015), agrees. She cites research stating that over 95% of adolescent children use the Internet, and more than 80% use social media. The typical teenager sends over 50 text messages a day. Eighty percent of teenagers play video games. By the time they're 21, the average young person (especially boys) will have played at least 10,000 hours of video games. This is the amount of time (according to Malcolm Gladwell in his book, *Outliers)* needed to become an expert in anything. According to Nutt, gamers score higher on tests for risk-taking. She reports that MRI scans of many adolescent brains show fewer connected circuits in the PFC, which explains their lower capacity for critical thinking.

How the Damage Is Done

Video games, social media, and smartphones have obvious benefits. But the makers of these highly profitable technologies consciously design them to make people want to use them more often. And like anything done to excess, they can do harm.

The difference between moderation and excess applies to almost anything. For example, occasionally eating pizza or apple pie can be both nutritious and delightful. But a steady diet of high-sugar foods and other carbohydrates can lead to obesity and diabetes.

Outdoor activities can return us to nature and be an opportunity for exercise. Also, exposure to sunshine promotes vitamin D production, which is vital to good health, as well as serotonin, which enhances mood. But too much exposure to the sun's rays can cause sunburn and skin cancer.

Even compassion can be overdone. In the extreme, a desire to be helpful can lead to giving unwanted advice, creating dependency, and neglecting one's own needs. Even drinking too much water can make a person ill, though in practice this is difficult to achieve.

As I've already stated, computers, smartphones and video games have many benefits. But like anything else, exposure to excessive screen time can harm a child's brain and behavior.

This is the subject of Dr. Victoria Dunckley's book, *Reset Your Child's Brain* (2015). Dunckley is a psychiatrist who has treated hundreds of young people for video game addiction and a new disorder she calls "electronic screen syndrome" (ESS). The problems arise when a child's nervous system is overly aroused by intense or frequent interactions with electronic media. A child's reaction to video games is particularly intense. The human brain responds to computer-generated images as if they were real. A modern video game's rapidly-changing fast-paced action content causes a stress-producing fight-or-flight reaction, which Dunckley refers to as "hyperarousal."

The brain triggers the release of adrenalin, which speeds up the heart. It also releases cortisol, which alters blood sugar level for a burst of energy. All this prepares the body for action. This happens whether what is sensed is real or something viewed on-screen. Once the real or virtual situation is successfully dealt with, the brain triggers the release of dopamine, the pleasure-producing brain chemical. This physical arousal, followed by a psychological reward, is what can make gaming addictive.

For our ancient ancestors, this stress reaction was momentary and necessary for survival. Your game-playing adolescent, however, is likely to sustain this experience over and over for hours, day after day, causing a condition of unrelieved, chronic stress. While adrenaline quickly subsides when the stimulus is removed, cortisol stays in the brain longer. Kids who play video games for hours will feel emotionally charged up long afterward. This is especially true of violent and first-person shooter games. Chronic stress is toxic because the persistent presence of cortisol in the brain destroys brain cells, particularly in the hippocampus, which is involved in memory and learning. This happens because cortisol binds to receptors inside many brain cells. With prolonged exposure to cortisol, brain cells can become overloaded with calcium, which causes them to fire too frequently and die. Too much residual cortisol also suppresses the immune system, which is vital to good health.

According to the report, "Excessive Stress Disrupts the Architecture of the Developing Brain" (National Scientific Council on the Developing Child, 2014), chronic, unrelieved stress can lead to "weakened bodily systems and brain architecture, with lifelong repercussions." It also states that during "sensitive periods of brain development, the regions of the brain involved in fear, anxiety, and impulsive responses may overproduce neural connections while those regions dedicated to reasoning, planning, and behavioral control may produce fewer neural connections." This is what can happen to kids who spend hours working a game console.

Immediate symptoms of chronic stress can include irritability, emotional melt-downs, poor self-regulation, low frustration tolerance, and disorganized behavior. In the longer term, a child may suffer sleep deprivation, increased sensitivity to stress, poor short-term memory, learning problems, weight gain, and social immaturity.

As brain scientists remind us, "the brain cells that fire together wire together." One of the worst impacts of hours spent playing intense video games is its huge "opportunity cost." Gaming engages physical and emotional reactivity. So instead of spending hours exercising critical thinking skills that develop the PFC, your child will be wiring his brain to react to fight-or-flight situations. And while they engage in this entertaining activity, the prefrontal cortex relentlessly continues to prune away unused connections, which can eventually reduce intellectual capacity—the equivalent of permanent damage in the part of the brain that makes a person smart.

According to Dr. Dunckley, the problem isn't just video games. Her years of experience dealing with the problems caused by excessive screen time have led her to conclude that "all screen activities provide unnatural stimulation to the nervous system." For example, when a young person engages with social media, adrenalin and cortisol provide mild jolts of excitement and anticipation. Later, when she hears the tone signaling a text message, is friended on Facebook, or receives a comment on a photo via Instagram, this reward triggers her brain to release dopamine, producing a feeling of pleasure, which keeps her coming back for more.

Unlike the intense experience of playing video games, the effects of exposure to milder media slowly build up over time. Dunckley says that the impact of many screen devices can accumulate: computers, televisions, non-violent video games, smartphones, iPads, tablets, laptops—even digital cameras, watches, and e-readers, regardless of how they're used, such as learning on educational programs, texting, chatting, instant messaging, gaming, emailing, social media and surfing the Internet.

This dynamic can become a negative feedback loop. Kids feel stress from schoolwork, activities, and parental expectations, so they use gaming and social networking to relieve the stress; doing this becomes addictive, wasting time and neglecting in-person relationships; the reduced time left for schoolwork and activities increases stress; which causes the cycle to continue, spiraling downward.

The Impact on Kids' Brains

Teens are especially at risk because the judgment part of their brain is under construction. This makes it hard for them to grasp and accept the consequences of excessive screen time, limit their use of the devices, and make healthy life decisions. They may not agree that connecting mostly through social media will make it harder to build vital social skills, or that time spent accessing electronic entertainment can subvert reading, writing and other key learning skills. And worse: *instead of wiring their PFC for reasoning, judgment and critical thinking, the devices are training their brains to react impulsively and emotionally.*

I once spent several days at the beach with family. One of the young people in attendance was Travis, a 17-year-old boy. I enjoy talking with teenagers, and I look forward to spending time with them. I found Travis in the family room playing a video game on his laptop. I noticed a battle tank on the screen. I asked him what game he was playing, but he didn't answer. I assumed he was so absorbed he didn't realize I was talking to him. I put my hand on his shoulder and asked again.

"It's just a game," he mumbled.

"Looks like fun."

"Mmm."

Whenever I saw him that weekend he was either by himself playing the game or watching TV. I rarely saw him with other members of the family or out on the beach. This was several years ago, before I read Dr. Dunckley's book and learned more about the deleterious effects of excessive screen time.

In many ways, excessive exposure to video screens is a lot like consuming too much alcohol. Both activities:

- Are legal, accepted, and heavily promoted by our culture
- Can be beneficial when used in moderation
- Are often used excessively

- Reward the user with a pleasure-producing dopamine response
- Can be used to escape the problems and pressures of the real world
- When used excessively, can have a negative impact on health
- Cause the real world to seem less interesting or even boring by comparison
- Harm relationships and lead to poor social skills
- Can be habit-forming and lead to addiction
- Have a negative effect on judgment
- Do physical damage to specific brain areas
- Interfere with PFC development (critical thinking skills)

Parents and teens alike are dazzled by the new function-rich devices that arrive each year. I recently spoke to a mother of three boys, aged 11, 13 and 15. She told me her ex-husband takes an active interest in their sons, and recently he gave his oldest boy a smartphone for his birthday. The boy had talked about wanting one for a long time, and he was thrilled. The father also gave all three sons video game consoles for Christmas. He won a lot of "best dad" points with these gifts. She said she and the father felt that time spent playing interactive video games would be time spent off the street and out of trouble. Also, the games would boost their kids' popularity with friends, and surely the boys would acquire mental skills they wouldn't acquire from passively watching television. These parents are typical of moms and dads who love their kids but haven't learned about the downsides of excessive screen exposure.

Your child will want one of these devices, too, and you'll need to discuss the issues with him. As in most things, moderation is the key to enjoying the benefits without your teen becoming a victim of the downsides of excessive exposure. And with these alluring technologies, the slope to addiction is slippery. Establishing boundaries and limits to protect your child's developing PFC won't be easy. Adolescents have a hard time imagining the future and foreseeing consequences, so explaining the harmful effects of excessive screen time will be a daunting challenge.

Keep in mind that these amazing 21st century devices have distinct benefits. Problems arise, as with most things, when moderate use becomes overuse. For example, one goal should be to balance involvement in social media with forming and growing in-person relationships. It's important that virtual experiences don't replace real, natural ones. In the end, protecting your child's brain is vital to developing a superior mind.

Major takeaways from Chapter 6:

- The adverse impact of too much screen time has been a problem since the invention of television.
- In the 21st century, the introduction of many new screen-based devices has made it possible for teens to accumulate enough screen time to interrupt normal growth of the PFC.
- The technology with the greatest potential for addiction and adverse effect on the brain is shooter video games, which happen to be the most popular (and profitable) technologies.

- The accumulated effect of long-term exposure to milder media can also affect the brain.
- Time spent watching video screens is time not spent on things that prepare a kid for life: being with friends, learning life skills, playing outdoors, reading books, getting involved in community service, or participating in family activities.

Recommended actions for parents:

- Recognize what screen addiction looks like. Everything may start out fine; but like alcohol or drug use, it's a slippery slope. If your child chooses to spend time gaming or using a smartphone instead of physical activities, homework, family responsibilities, getting enough sleep, being with friends, or participating in art, music or reading, he or she is probably addicted.
- Share what you've learned. Explain that while these devices can be beneficial, overuse can create serious health and behavior problems, and it can disrupt brain development. If talking to your teen about this seems daunting, have him read this chapter, then discuss it with him.
- If excessive screen time has already led to behavioral issues, consider Dr. Victoria Dunckley's "Reset." Her clinically proven remedy is the equivalent of "cold turkey." It involves a week of discussion, planning and preparation, followed by three weeks of total abstinence: all screens, including television. The Reset also involves substituting screen time with "green time," i.e., time in the real world doing constructive activities with people, often outdoors. The period of abstinence can be followed

by limited, strictly managed screen exposure. As you can imagine, teens who are already addicted users could greet this program with anger, defiance, and worse. Her recommendations for making the Reset work are outlined in her book, *Reset Your Child's Brain*. Parents seeking more information about the dangers of excessive screen time can consult her extensive chapter endnotes.

- Encourage real-world alternatives. The goal is balance. Spending too much time with video games can harm the brain, and doing so takes time away from activities that help a child grow up, such as reading, board games, playing a musical instrument, sports, outdoor activities, family outings, construction projects, community service, or money-making enterprises.
- Make kids aware of online dangers. Talk about cyberbullying, sexting, pornography, and identity theft with your kids.

BONUS: Go to https://DrDennyCoates.com/bonus to download a free guide: "A Practical Plan to Moderate Teen Screen Exposure."

Protecting your adolescent's brain is crucial, because it allows the many activities that stimulate PFC growth to have a positive impact. In Chapter 8 of Part Three you'll learn about quite a few such activities, many of which are likely to appeal to your teenager.

Part Three

WHAT TEENS CAN DO

The three chapters of Part Three are written for you, the adolescent who has the potential to mature into a happy, successful, independent adult.

The all-important question is, after learning about the power of the PFC and what you have to do to develop it: *Will you actually do the work?*

Because no one can do it for you.

Chapter 7 defines this responsibility and asks you to decide if you feel having a superior intellect is something you want enough to make the effort.

Chapter 8 describes the many academic courses, extracurricular activities, and games that can help you wire the smart part of your brain.

Chapter 9 explains how drinking alcohol and using drugs during adolescence can significantly disrupt the development of the smart part of your brain. The purpose of connecting the dots about substance abuse is to encourage you to say no to drugs and defer even moderate alcohol use until you're a fully grown adult.

7

Take Charge of Your Growing Brain

"The sum of one's intelligence is the sum of one's habits of mind."

– Lauren Resnick –
American educational psychologist (1936—)

"You've got to do your own growing, no matter how tall your grandfather was."

– Irish Proverb –

I wrote the first six chapters for adult readers, though I've encouraged them to share it with young people. In this chapter, however, and the two that follow it, I address you—the teen—directly, because while you can, indeed, grow a smarter brain, no one can do this for you.

First, a few words about your world, because sometimes in all the fun and friendship, it's easy to forget where all this is headed.

There are all kinds of kids in every high school. You've got your jocks, your bookworms, your cool kids, your bullies and outlaws, and maybe even goths, who, like everyone else, are trying to express their individuality.

But one thing about high school culture: it only lasts four years. This book is about preparing for what you'll make happen *the rest of your life*. I know your mind is occupied with schoolwork and activities with friends, so the idea of your future can seem vague and distant. But believe me, it's really going to happen and will arrive faster than you can imagine. It will bring both opportunities and challenges. So you don't want to wait until high school is almost over to start caring about what's coming next.

Your chance to prepare is right now.

After graduation, something unexpected happens. All the kids you were close to will go their separate ways. They'll make one big decision after another and head down new paths. Some will enter the military service. Many will go to college. Some might get married right away. Some may decide to work entry-level jobs in town. And some will continue to live at home, playing games, getting high, and avoiding work whenever possible. A few could become drug addicts, suffer mental illness, or die untimely deaths. One of my best friends in high school died in a car wreck while driving back to his dorm from a college drinking party.

In most cases, you won't know how their lives turned out until years later at a high school reunion—if they show up. Say you're 33 years old and you do show up. You'll be surprised to find out that some of the most unlikely of your former classmates have become quite successful: business owners, scientists, lawyers, doctors, engineers, bankers, etc.

These are the folks who, when they were young, happened to be building a smarter brain, whether they realized it at the time or not.

Others will not have changed one bit. Maybe they're still living in the town they grew up in.

It's kind of amazing, but this happens to every high school class.

And you—what opportunities await you after high school? Will you surprise your classmates by the paths you choose to follow? Will you work hard and maybe along the way catch a little good luck?

Right now, during your adolescent years, you can wire your brain for powerful thinking skills, even if you have no idea that you're doing so, if you're fortunate enough to be encouraged and mentored by adults who care about you; if you get involved in the right hobbies, extracurricular activities, courses and games; and if you take them seriously and apply yourself.

How cool is that?

As I shared in Chapter 3, this is what happened to me. Ignorant of how my adolescent brain was growing, I simply got lucky in a lot of ways. But this kind of luck doesn't happen for everybody. The stories of my seven brothers and sisters were nothing like mine.

Hard work, with a little good luck mixed in, has been the system for achieving a fine mind for thousands of years. Certain young people, unaware of how a growing brain becomes smart, happened to be encouraged and happened to do the kinds of things that wired their brain for superior intelligence.

But all too often, this doesn't happen.

I wrote this book to help you break away from this system, so you and your parents would know how your brain's prefrontal cortex (PFC) gets wired, how high the stakes are, and what you need to do to achieve the best result.

My goal for you and as many of your friends as possible is to:

Take luck out of the equation.

A recurring theme in nearly every chapter of this book: you really can build a foundation for an impressive intellect, *if you do the work*. It's your brain, so only you can call the shots. No one can make you learn. And no one can keep you from learning, if that's what you want to do.

In the next chapter I explain what you can do that will make a difference. You'll have many activities to choose from. It's like building physical fitness: there are lots of ways to go about it. But if you don't work at it, nothing happens. Your brain won't automatically wire itself just because it's time. *And no one can do it for you.*

If you could get smart without making an effort, then everybody would grow up brilliant.

And you know that's not the case. A whole lot of people become adults with average or below-average mental capacity. A whole lot of people struggle to get traction in life.

I assume someone who cares about you gave you this book, and you've already read the first six chapters. If so, then you know you don't have to get lucky. You're among the first generation in human history who can learn how the smart part of the brain gets smart. You can now make conscious choices to get involved in things that will wire your PFC for some really powerful thinking skills, the kind of executive and critical thinking skills that will make it easier for you to deal with life's challenges and set you up to succeed in the life you choose.

In other words, *you can make it happen.*

Do you want it to happen? Do you care about being smart? Do you care about what's possible for you as an adult?

I ask, because not every teenager does.

Maybe you haven't given these questions much thought. But now that you've read this far, maybe you realize that you do care. Maybe you do want a smarter brain. Maybe you really do want it badly enough to do something about it.

As you learned in Chapter 1, there are quite a few important thinking skills. These aren't simple skills that you can master in a few tries. To wire your brain for these skills you'll have to practice them diligently for years.

Acquiring thinking skills is like mastering the skills for a sport, such as tennis. Learning to hit a backhand shot is fun, but getting good at it isn't easy. You have to put in lots of practice. You gotta do the reps, as they say. You have to work at it consistently over time.

But in one way, building a smarter brain isn't like getting good at a sport. The progress you'll be making in your brain will be slow, silent, and invisible. In other words, while you're steadily creating neural pathways for thinking skills, you won't be able to gauge how well you're doing.

This isn't something you can put off until later when you're an adult. When you were younger, your brain had periods of readiness for learning basic skills like seeing, sitting, standing, walking and talking. Brain cells in the parts of the brain that handle these skills were ready to be connected into circuits. You eagerly did the work, even after trying and failing dozens of times, until you mastered the skills. And then the cells that were still unconnected in that part of the brain were eliminated.

The last area to construct foundation circuits is your PFC, and its period of readiness is *right now*—the growing up years we call adolescence. The window of opportunity is open, but in a few years it will close—after the unused cells in your PFC have been eliminated.

Your judgment and problem solving abilities are likely to improve while you're still growing up, which is great. But the biggest payoff will come in the future after this foundation is set and you're ready to build on it. Hopefully the foundation you end up with will be massive.

Also, you'll need to protect your brain from certain aspects of your culture, such as alcohol, drugs, and excessive screen time, which can introduce chemicals into your brain that derail normal development.

So read on and finish the book. The next chapter will focus on the activities that will give you the biggest payoff. Hopefully some of the stuff you're already doing is the right stuff, and you can give it more emphasis. Or you'll learn about new things that are fun and will help you grow a smarter brain.

And at the end of the book you can decide if you want to make the kind of commitment it will take to do this for yourself while you still can.

Major takeaways from Chapter 7:

- You can consciously and deliberately grow a smarter brain.
- It won't just happen. You'll have to exercise thinking skills throughout your teen years in order to wire the circuits in your PFC.
- No one can make you do the work, and no one can do it for you.

Recommended actions for teens:

- Decide how important it is to you to grow a smarter brain.
- If it means a lot to you, make a commitment to get involved in courses, activities, and games that will help.

- When your parents or other adults ask for your thoughts and opinions, take their requests seriously and give your best answers.
- Check out the options in Chapter 8, and decide which ones appeal to you most.
- After you read Chapter 9, make a commitment to protect your still-developing brain.

8

Get Involved in Courses, Activities, and Games That Make You Think

"The development of general ability for independent thinking and judgment should always be placed foremost, not the acquisition of special knowledge."

– Albert Einstein –
American physicist (1879-1955)

"That which is used develops, that which is not used wastes away."

– Hippocrates –
Greek physician (460-370 B.C.)

Thinking skills are like any other skills or habits. Any mental or physical action, if you do it over and over, will eventually cause the specific brain cells involved in the action to form a permanent circuit that triggers the behavior. This allows you to perform the behavior without deciding to do it or even thinking about it as you do it.

In other words, you can develop impressive thinking capacity simply by repeatedly doing things that involve the skills.

It's possible that some of the stuff you're doing already is helping you grow your brainpower. And yeah, maybe some of it isn't, such as:

- Texting and taking "selfies" and posting them on social media
- Listening and dancing to pop music
- Attending concerts
- Shopping
- Grooming
- Hanging out with friends
- Dating
- Reading comic books or romance novels
- Watching sitcoms on TV
- Playing video games

Don't get me wrong. You should enjoy your teen years. You don't have to stop doing these things. It's all about balance. But if being popular and having fun is the biggest thing in your life, there might not be much time left over for activities that grow a smarter brain.

School Courses That Can Develop Thinking Skills

A huge ah-ha: You can grow a smarter brain just by going to class. But not every class helps, and just showing up won't cut it.

The all-important questions: *What courses can you take that will best develop an amazing intellect? Do some courses have a greater impact on the PFC than others?*

I'll say this up front: most school courses can, to a degree, help you develop certain thinking skills, if:

- You take the course seriously and work hard to master the content
- You get inquisitive and learn more about the topic outside the classroom
- Your teacher explains "the why" along with "the what"

These are big ifs. Many kids just show up and do the minimums. If you daydream when you should be paying attention, blow off homework, cram for tests, or barely make a passing grade, you may get credit for the course, but your attendance probably won't do much to exercise your PFC.

Also, some courses may have huge practical value, such as a typing/keyboarding course, but do little to build critical thinking skills. Other courses are both valuable and fun, such as P.E. You should participate enthusiastically and get maximum benefit, but they aren't designed to build brainpower.

Many courses focus heavily on facts, such as history and government, which can be interesting and useful; but by themselves, facts don't engage the PFC. A lot depends on how a course is taught—*and whether you take the initiative to ask about why and how.* The willingness of your instructor to teach the why behind the facts can help your brain grow neural pathways in your PFC:

- You can learn to follow a recipe precisely and get an A in family and consumer science, but if you also learn why the recipes work and the chemistry of cooking, you develop brainpower.
- In shop you can learn to use tools, but if you design what you make, and find out how and why the tools work, you develop brainpower.
- You can learn to draw or paint, but if your instructor also explains why and how certain techniques work, you develop brainpower.
- You can practice a musical instrument and learn to perform difficult pieces, but if your music director also helps you understand what the composer was trying to express, you develop brainpower.
- What you learn about health and nutrition can prolong your life, if you apply it; but if you also learn how and why these practices benefit you, you develop brainpower.

On the other hand, some courses focus so much on why and how that it's nearly impossible to get a good grade without growing your PFC. For example, any course in:

- Science
- Technology

- Engineering
- Math
- Philosophy
- Essay writing
- Debate or Speech

Don't be intimidated by these courses. And don't sign up for them as if they're a kind of "medicine" that'll make you smart. If you make an effort, stay curious, and work hard, you can not only master the subject matter and grow your intellectual abilities, you can enjoy a really neat journey of discovery.

All science courses are about how the natural world works: how the earth has changed over billions of years, how weather works, how biological reproduction works, how you inherit certain genetic features, how scientists learn this stuff—all very cool, and learning it will grow your PFC.

While science helps us discover through testing, observing and measuring, philosophy helps us learn using logic. Probing for answers to hard questions will—if you follow along and do the thinking—cause your reasoning circuits to form.

Essay writing is about how to state a convincing case in writing. Debate is about stating a convincing case verbally. Speech is about telling a cogent story. The activities in these courses require you to reason and will wire your PFC.

Math is the language of science and engineering. You need math to figure out what the data mean. I recommend courses in algebra, geometry, trigonometry, calculus, and statistics. As you learn a procedure, always ask: *how will this help me solve problems in real life?*

I'm often asked which course is at the top of the list, the absolute best course for wiring the PFC. As I've said, a lot depends on your effort and how the course is taught. But my favorite course for exercising critical thinking is computer programming.

Every computer software program is a series of instructions, and every instruction has a logical purpose. Every line of code you write requires reasoning and cause-and-effect thinking. I know, some kids like to make fun of students who know how to program computers, referring to them as "geeks" or "nerds." This is an expression of the name-callers' own inadequacies and their envy of people who are smarter than they are.

So don't let their attitude discourage you. Take the course. You'll discover how much fun it is to create a program that tells a computer what to do.

Both my sons developed a passionate interest in computer programming, and today are successful information technology experts. I sometimes joke that both of them are smarter than I am. But in truth, it's not a joke. As an older guy who has thought about the important things in life for decades, I may have them beat in the wisdom department. But they both have amazing intellects, and in some ways they are, indeed, smarter than I am. And they both make more money than I do.

Many years ago several colleagues and I formed a team to present a weeklong course in creative problem solving. The night before the course began we sat in the hotel lobby enjoying drinks and conversation. At one point, someone asked, "Who was your favorite college professor?" It was fascinating to hear everyone's responses, and I never forgot one woman's answer:

"My economics professor. He taught me how to think, and it changed my life."

At the time, something about her story surprised me. *An economics course teaches economics. How can it help you learn to think?* I wouldn't know the full answer until a couple decades later, when I became immersed in the science of how the brain develops, learns and thinks.

My study of adolescent brain development has caused me to reflect on my own high school and college experience. As a youth, I was able to make good grades without thinking deeply about what I was learning, and I had no mentors who made me think. My passions were writing poetry and playing golf. I was at the golf course every day, weather permitting. As I recall my high school years, I realize I did very little to construct a foundation for critical thinking and judgment. Nevertheless, I graduated with a 4.0 GPA and was my class valedictorian.

At West Point, academics were a lot more challenging. All classes were mandatory—over 20 credit hours per semester. I even attended classes on Saturday morning. Even though I began my studies with a lot of confidence, I was surprised to discover that many of my classmates were a lot smarter than I. They seemed to grasp difficult subject matter effortlessly, while I struggled. Despite my best efforts, at the end of freshman year I was barely in the top 25% of my class. It was humbling. I wasn't used to this feeling, and it motivated me to work harder.

Looking back, I realize that the greatest value of my West Point education was that the courses taught me how to think. I didn't know it at the time, but all those courses in mathematics and engineering during my late adolescence drove me to exercise my PFC for hours every day. Military history was about analyzing why battles were won and lost. Even English classes pushed me to understand literature, not simply to enjoy it. So for me, it wasn't one professor who taught me how to think; it was the whole curriculum.

By the time I was a senior, my semester grades were in the top 5% of my class. I worked hard to achieve that; I had always been, after all, a motivated student. But I often wonder what it would have been like if I had arrived at West Point better prepared intellectually.

Extracurricular Activities That Grow the Prefrontal Cortex

Not every school system offers courses such as philosophy, debate, and computer programming, which require students to think logically. But often there's enough interest among teachers and students to form school-sponsored clubs related to these topics. All extracurricular activities are fun, but not all of them have the side-benefit of growing a smarter brain. Examples of extracurricular activities that stimulate the PFC to wire itself:

- Science-centered clubs focused on animal husbandry, biology, chemistry, robotics, astronomy, rocketry, etc.
- Technology-centered clubs focused on electronics, web design, computer programming, etc.
- Student leadership
- School newspaper and yearbook
- Event planning committees

- Competitive teams for debate, math, science, robotics, etc.
- Clubs for board games such as chess, go, and Dungeons and Dragons

Beyond school, some communities sponsor youth organizations that include opportunities for service and leadership development, which exercise planning and organization skills. Examples:

- 4-H
- Future Farmers of America
- Scouting
- Big Brother Big Sister
- Youth camps and retreats

An essential ingredient in your participation should be a genuine interest in the activity, although you may not be sure about your level of interest until you get involved. If you discover that it isn't what you expected, you can try something else.

Some mind-engaging activities can be self-directed, such as a hobby or a personal interest. For example, teenagers who start money-making ventures will be involved in planning, organizing, and managing their business—all of which stimulate activity in the PFC.

There are quite a few fun activities that require little or no critical thinking. But this doesn't mean you should avoid them, because they're likely to have other benefits. For example, team sports are fun, create friendships, promote physical fitness, develop personal strengths, and exercise teamwork and social skills. Art- and craft-related instruction can develop basic art skills, exercise creativity, and build self-esteem. Service organizations can teach the value of making a contribution to the community and give a chance to exercise leadership.

Games That Build Brainpower

The point of a game is to have fun, right? The cool thing is, some games are not only fun, they make you think. These are the games that wire your brain for thinking skills.

The No. 1 most powerful brain-builder game is chess.

Some people think you have to be smart to play chess. That's a misleading statement, because anybody can learn to play, and if you continue to play chess, it will make you smarter. The reason is that in order to win at chess, you have to learn to foresee the consequences of your own moves and the possible moves of your opponent.

Chess is considered a two-player strategy board game because you win by out-thinking your opponent. If you've never played chess before, two excellent books can help you learn without a teacher: *Chess for Kids* (2006), by Michael Basman, and *One Move at a Time: How to Play and Win at Chess...and Life!* (2007), by Orrin C. Hudson. The second book is unique because Part 2 describes 20 life lessons you can learn from playing chess.

People have enjoyed chess for over a thousand years. If you join a school chess team, you can not only learn advanced strategies, you can earn high-level status and even qualify to play in interschool, national, and international championships.

Other ancient and still-popular strategy board games include go, backgammon, checkers, scrabble, and dominoes. Yes, dominoes! Many years ago on a business trip to Mexico, I played dominoes with some experienced players. They beat me every time because of their advanced knowledge of strategies. All these games can be played online, though for me it's more fun to sit down with another player in person.

My two sons learned to play chess, but they also started playing another multi-player strategy game: Dungeons and Dragons. Over 30 years later my oldest son still meets with a D&D group every week.

Both boys grew up with video games. But back in the 1980s, the Commodore 64, one of the first PCs, supported only simple games, like Pac-Man, on a small black-and-white screen.

Now, in the 21st century, the video game industry is a much bigger business than the motion picture industry. There are thousands of video games to choose from. It's a highly innovative technology that is moving towards virtual reality games.

Many of today's video games are strategy games, but many others are just fun and exciting games that do nothing to stimulate the PFC. Some of these are harmless entertainments, such as those that simulate being in a race car, flying a jet plane, and competing in sports.

There are also educational games that exercise perception, memory, vocabulary, and facts about science and history.

And there are the action games, which simulate violence. The games that put the player in the role of shooter are particularly intense. As I described in Chapter 6, the brain does not discriminate between real action and screen-based action. Playing the game causes the release of adrenaline and cortisol, which provides the sensation of excitement and stress. Winning a battle and earning a higher level of competition causes the release of dopamine, the pleasure/reward brain chemical. Because of this, all video games are potentially addictive; but violent games, because of their intensity, involvement, and rewards, are more addictive.

The residual cortisol that collects in your brain from playing violent action games can cause cell damage and retard your efforts to grow a smarter brain. Like using alcohol or drugs during adolescence, it can cause a permanent degradation of intellectual capacity.

My recommendation: resist the invitations of your friends to participate in action shooter games such as Fortnite, Call of Duty, Halo, GoldenEye, and Doom. Yes, they're exciting. But like other activities kids get involved in, your friends may not appreciate the destructive side effects.

On the other hand, getting hooked on a game that builds your brain is a healthy addiction, like being addicted to a "runner's high" while jogging.

Stick with the countless other entertaining video console or online games, especially challenging strategy games such as: Age of Empires, Age of Mythology, StarCraft, Stellaris, Off-World Trading Company, Civilization, Portal, Minecraft, and SimCity. Some are exciting war-themed games, such as

Command & Conquer, Company of Heroes, Total War, World of Warships, Throne, Vikings, and Stormfall. But these show the action from a "2000-foot" perspective, so they're a lot like chess.

What I've outlined in this chapter is a triple-whammy of powerful, fun ways to build impressive brainpower during your teen years: courses, extra-curricular activities, and games that will steadily wire your PFC for an impressive array of thinking skills.

My recommendation: stay involved in all three. You're going to attend classes, get involved in activities, and play games anyway. So why not choose some that will help you grow a smarter brain?

And while you're doing that, you'll need to protect your brain from foreign substances that can disrupt your best efforts. The next chapter explains why and how.

Major takeaways from Chapter 8:

- Some courses focus exclusively on skills and content that will exercise your PFC. The best of these involve computer programming.
- Nearly every course will exercise your PFC to some degree, provided you take it seriously and work hard at it.
- Quite a few extra-curricular activities build brainpower.
- The best games for exercising the PFC are strategy games.

- The most powerful brain-building game is chess.

Recommended actions for teens:

- Spend more time being involved in courses and activities that are known to make you think.
- Google the Internet to discover which strategy games appeal to you.
- Because of their potential for addiction and harm to your growing PFC, avoid violent action games, especially shooter games.
- Consider board strategy games, such as chess and Dungeons & Dragons. They not only make you think, you get the added benefit of real, in-person interaction with friends.

BONUS: Go to https://DrDennyCoates.com/bonus to download a free guide: "The 5 Secrets to Getting Better at Anything."

9

Protect Your Brain from Alcohol and Drugs

"Sometimes when I'm working with teens, I try to reason with them that if they're doing drugs or alcohol that evening, it may not just be affecting their brains for that night or that weekend, but for the next 80 years of their life."

– Jay Giedd, MD –
Pioneering adolescent brain researcher, NIMH

In the previous chapter you learned about the engaging school and community activities that can stimulate your prefrontal cortex (PFC) to connect circuits for powerful thinking skills.

But there's a potential problem. Two things in today's youth culture can mess with your brain chemistry while you're trying to ingrain new thinking skills:

- Excessive screen time
- Substance abuse

In Chapter 6, I explained how too much exposure to games, smartphones, computers and other screen-based technologies can disrupt your ability to develop your PFC. Part of the solution was moderation: using devices to enjoy the benefits, but avoiding excessive use.

For adults, moderate use of alcohol can help them relax at the end of a busy day and facilitate socializing with friends. However, because of the negative effect alcohol and drugs can have on your still-developing brain, moderation isn't the answer for teens.

This chapter reveals the permanent adverse effects using alcohol or drugs during adolescence can have on your PFC. I've written this chapter for you, because while parents can help you manage your screen time, they can't prevent you from consuming alcohol and drugs. Only you can make this choice.

If you consume the wrong substances in the wrong quantities at the wrong times, it can prevent the smart part of your brain from establishing the wiring you're working on. In a very real sense, this is the equivalent of permanent brain damage.

And yes, in the worst case, substance abuse can cause addiction and even kill you.

The Effect of Substance Abuse on the Adolescent Brain

I'll describe how this works by comparing the brain of a teenager to the brain of an infant in the mother's womb.

I recently visited a local hangout called "The Pour Haus." A large, mostly young crowd had gathered for happy hour, and several interesting craft beers were on tap. After half an hour of enjoying the festive atmosphere, I found the men's room. As I washed my hands, I noticed a small sign next to the mirror:

Drinking any kind of alcohol can hurt your baby's brain, heart, kidneys and other organs and cause birth defects. The safest choice is to not drink at all when you are pregnant or trying to become pregnant. If you think you might be pregnant, think before you drink.

While this is an important message, I was surprised to see it at eye level in the men's room. Since the establishment served alcohol, I assumed the purpose of the message was to protect the management from liability and that a similar sign was also posted in the ladies' restroom.

The warning is pretty much what nearly every pregnant woman hears from her doctor. The unborn baby's body and brain slowly develop during the nine months in the mother's womb, and the blood of the mother provides nutrients to the placenta, which nurtures the growing fetus. Therefore, if the mother consumes alcohol or drugs during pregnancy, the chemicals can make their way to the baby's underdeveloped body and brain and interfere with normal growth, causing permanent, life-limiting damage. Doctors refer to the range of possible birth defects as "fetal alcohol spectrum disorder," or FASD. The brain damage and visible body disfigurement can be shocking. The actual damage to the brain can't be seen, of course; but it can permanently affect the victim's behavior.

It's an effective warning; the thought of birth defects and brain damage is enough to give any expectant parent cold chills. Doctors' warnings prevent most of these horrors, but not all. In the U.S., over half a million children suffer from FASD. Many of them will never be able to live independently or have normal relationships.

So what does this have to do with healthy teenagers?

The answer:

By consuming alcohol or drugs, you can do this kind of damage to your developing brain all by yourself.

The damage doesn't happen because foreign substances destroy brain cells. This myth has been disproven. What really happens is that the presence of alcohol or drugs makes it hard for brain cells to connect, which can slow or even prevent the process of creating circuits. Depending on the frequency and magnitude of substance abuse—and how much you're actually exercising thinking skills, consumption may not stop circuits from forming completely. The impact of these variables on the growing PFC isn't well understood. What's more likely to happen is a delay in your development and a diminishment of your capacity to handle challenging mental tasks.

In addition to alcohol, marijuana is another drug that can inhibit adolescent PFC development. Teens have always experimented with marijuana, and today's generation can point to the fact that marijuana is now legal in many states and has quite a few medical uses. Also, marijuana doesn't contain nicotine; its high comes from the psychoactive substance called tetrahydrocannabinol (THC). All of which can lead a young person to conclude that while cigarettes may be dangerous to their health, "weed" is okay.

Don't believe it.

According to Dr. Susan Weiss, director of the division of research at the National Institute on Drug Abuse (NIDA), "There's a growing literature, and it's all pointing in the same direction: starting young and using frequently may disrupt brain development." Referring to a Duke University study that followed over 1,000 New Zealanders born in 1972, the team found that persistent marijuana use during adolescence caused a drop in brainpower equivalent to about six IQ points. "That's in the same realm as what you'd see with lead exposure."

The University of Barcelona reviewed 43 studies of chronic cannabis use. Eight of the studies, such as "Marijuana and the Developing Brain" (*American Psychological Association*, November 2015, Vol. 46, No. 10) suggested that both structural and functional brain changes emerge soon after adolescents start using, and the changes are still evident after a month of abstaining from the drug. Some of the brain abnormalities have been linked to cognitive differences. Regular, heavy marijuana consumers—those who reported smoking it five of the last seven days, and more than 2,500 times in their lives—had damage to their brains' white matter, which enables communication among neurons. Marijuana users also had changes in shape and volume in memory and emotion brain areas. Adults who started smoking during their teens ended up using twice as often as those who picked up the habit as adults.

So you could make the mistake of going along with the fun, binge-drinking at parties and experimenting with drugs. But the chances you take are not trivial: if certain thinking skill circuits have trouble forming, you could potentially suffer a *life-long diminishment of your ability to focus, reason, solve problems, and use good judgment.*

Why do teens still take these risks?

Simple: either they haven't been told about the recent research into the still-developing adolescent brain, or they've discounted the warnings and don't appreciate the consequences.

But you do.

When it comes to the future of your growing brain, consuming alcohol and drugs are clearly not worth the risk. It would be as if an artist carelessly whacked away at a block of marble while trying to sculpt a masterpiece. If you abuse alcohol or drugs, your work to develop a superior mind may not have the desired effect.

My guess is that as I write this book, very few young people in middle school, high school or even college understand the risks they take when they drink alcohol or use drugs.

A friend of mine told me about his 27-year-old nephew, who as a teenager smoked marijuana and drank a lot. He said the young man still lives at home and acts and talks the same way he did when he was 15.

One of my colleagues, who helps run an addiction recovery service, told me he began drinking at the age of twelve and smoked a lot of grass. He said that when he grew up these habits had affected his ability to manage his life and be effective at work. He's worked hard to compensate for it ever since, but progress has been limited.

Personally, I dodged this bullet because I happened to be raised in a religious family where using substances was forbidden.

The simple, essential (but not always so easy) solution:

Avoid using illegal drugs altogether; wait until after adolescence (the mid-twenties) to begin drinking alcohol.

Over-the-Counter and Prescription Drugs

You can get high and risk brain damage with legal drugs, too. Lots of kids do.

Dextromethorphan (DXM) is the active ingredient in more than 100 over-the-counter (OTC) cough medicines and is the most widely used cough suppressant ingredient in the United States. When taken as directed, medicines that contain DXM are safe and effective. However, when taken excessively, DXM can produce dangerous side effects.

While millions of Americans rely on OTC cough medicines for relief, about one out of 30 teens reports abusing it. To get "high," teens often take more than 25 times the recommended dose of these medicines.

The bottom line: whether drugs are legal or illegal, prescription or OTC, the consequences of abuse can go beyond drunkenness, risk-taking behavior, and addiction. If you inhibit the thinking part of your brain when it's trying to grow, you could experience a life-long diminishment in brainpower.

Tobacco and Vaping

Recent research, reviewed on the website of the National Institutes of Health, showed that nicotine acts directly on the PFC and produces "neurotoxic effects" that can interrupt normal development. According to Natalia A. Goriounova and Huibert D. Mansvelder in "Short- and Long-Term Consequences of Nicotine Exposure during Adolescence for Prefrontal Cortex Neuronal Network Function" (*Cold Spring Harbor Perspectives in Medicine*, 2012), smoking during adolescence is associated with disturbances in working memory, attention, and PFC activation. Other studies show that teen smoking can cause academic problems and can set up a young person for mental and behavioral limitations later as adults.

The media sometimes portrays smoking as cool and a sign of maturity. Today, electronic cigarettes are considered even cooler, especially among teens. Kids often believe e-cigs are safer than tobacco cigarettes. But all electronic cigarettes, which include e-pens, e-pipes, e-hookah, and e-cigars, are essentially nicotine delivery systems. The main liquid component consists of nicotine extracted from tobacco and mixed with a base (usually propylene glycol), and may also include flavorings, colorings and other chemicals. In fact, e-cigs leave two or three times as much tar in the lungs as regular cigarettes.

Despite what some of your peers believe, smoking isn't a sign of "adult behavior." Most adults don't smoke. Why? They don't want to get lung cancer. They know that just breathing someone else's smoke can give them lung cancer.

Adolescents are more vulnerable to nicotine addiction than adults. Seventy percent say they've tried smoking, largely because of peer pressure, and 20% of teens say they're addicted. This is because teens:

- Become addicted more quickly than adults do
- Feel more intense rewards from brain pleasure centers
- Tend to underestimate the risks
- Have less impulse control than adults
- Are more influenced by peer pressure

Teens who think smoking is cool choose not to acknowledge the risks. What about you? The decision to smoke reminds me of that scene in the movie "Dirty Harry," when detective Hallahan, played by Clint Eastwood, points a .44 magnum at the villain and dares him to reach for his gun: "You've gotta ask yourself one question: Do I feel lucky?"

Major takeaways from Chapter 9:

- Using alcohol, nicotine, or drugs before your early-to-mid 20s, while your PFC is still developing, can permanently limit your brainpower.

Recommended actions for teens:

- Take responsibility for your own body and brain. Your parents may encourage abstinence, but you're the one who has to say no.
- Protect your PFC. Avoid experimenting with alcohol or drugs during your adolescent years. Since no one knows for sure exactly when or how much abuse will have a negative impact on your developing brain, never abuse drugs—legal or illegal—and wait until you're a fully grown adult to drink (and then do so moderately).
- Choose friends who aren't using alcohol or drugs.

CONCLUSION

Because this book has been written for both parents and teens, I include a summary chapter for both.

10

For Teens: It's Your Brain and Your Life

"Being a teen today is tougher than ever. While your grandparents may have had to walk uphill to school in the snow, you have a different set of challenges to navigate: like media overload, party drugs, internet porn, date rape, terrorism, global competition, depression, and heavy peer pressure."

– Sean Covey –
American business executive and author of *The 6 Most Important Decisions You'll Ever Make* (1964—)

If you're fortunate, being a growing teenager means being a part of a family. But the older you get, the more you see yourself as a separate individual, learning about the world, thinking about your place in it, and spending more time with friends who are doing the same thing.

Someday—sooner than you think—your time living with your family will be over. You'll be an adult, working and trying to create your own life.

Maybe imagining what this will be like is a little scary. And for good reason. Unlike your family, the world out there isn't set up to satisfy your needs and wants. And in that world, you'll be competing with other people—young and old—for the chance to achieve your goals and create the life you want.

No, it's not going to be easy. It's going to be hard. Your parents probably haven't told you how challenging it is to be an adult, because they want you to stay positive and optimistic.

But in spite of the stuff life will throw at you on a regular basis, you can succeed anyway, if you've prepared for the time when you leave home. I may have said this before, but allow me to say it again:

The purpose of adolescence is to prepare for the challenges of adult life.

Knowledge, skills, experience, strong character, and an outstanding mind—these are the things that will help you accomplish what you want. Adolescence is the time for you to start building these capacities as much as you can. Hopefully you'll continue to expand on who you are throughout your life. But one thing you'll discover as a young adult: some of your peers will be farther down the road than you. You'll be trying to play "catch-up."

Some of your classmates don't get this. Look around you and you'll see that they're focused mostly on what's happening now, not the future. They don't realize it, but they—like you—are heroes in their own stories. Hopefully their stories will have a happy ending, with good jobs, meaningful relationships, and a fulfilling life.

But like any hero, to achieve this victory, you—and they—will have to confront a gauntlet of dangers and challenges. Not dragons or evil villains like Darth Vader, but things like addiction to gaming or social networking; stress, anxiety and depression; the consequences of risk-taking; experimenting with alcohol and drugs; and other things your peers may urge you to get involved in.

If you work on wiring your prefrontal cortex, it will be a lot easier to use good judgment and avoid this gauntlet of pitfalls. It will help you travel your adolescent journey while avoiding all these mishaps.

And you may have a guide or two, like Yoda, who mentors Luke Skywalker in "Star Wars," or Harry Potter's Professor Dumbledore, or James Bond's Q, or Carlos Castaneda's Don Juan—or my grandfather. A coach or a teacher and even this book, which has been about how to train your mind to master an impressive array of thinking skills, can be one of your guides. It will also help if you stay humble and receptive to the wisdom of adults who care about you. If you work hard at it, it's reasonable to expect that by the time you leave home, your mind will be what people think of as "brilliant."

The big payoff comes during prime time: your life as an adult. If you really want to achieve success and get what you want, you'll need as much brainpower as you can acquire to compete with other adults who want the same things. As the hero in your own story, you can win. You can prevail.

In the end, it's your brain and your life. It's up to you to decide what's important to you and whether you care enough to make a commitment to growing a smarter brain. As one of your guides, *I'm calling you to action:*

- Take responsibility for growing your PFC.
- Get involved in courses, activities and games that make you think.
- Protect your brain from alcohol and drugs—and excessive screen time.

May the force be with you.

Major takeaways from Chapter 10.

- You are the hero of your own story.
- Every hero needs a guide, but not every hero heeds the advice.
- You'll achieve success at the end of your adolescent journey if you make the right choices and do the work.
- The world doesn't owe you anything. To get what you want in life, you'll have to earn everything.

Recommended actions for teens:

- Make a commitment to growing and protecting the smart part of your brain.
- As the hero of your own story, accept the wisdom and calls to action from people who want you to succeed.
- Hang out with friends who are preparing for success as an adult, rather than getting involved in things some teenagers mistakenly think are adult-like, such as using

drugs or alcohol; casual sex; and other forms of risk-taking.

11

For Parents: Preparing for Adolescence

"Don't worry that children never listen to you; worry that they are always watching you."

– Robert Fulghum –
American author (1937—)

To the above quote, I'll add my own summary: After puberty, young people mostly raise themselves. But since they're poorly equipped to do this, they need parents and other adults to coach them with love, guidance, encouragement, and support until they can make their way in the world.

It helps to keep in mind the purpose of adolescence—to prepare a young person to become a happy, responsible, effective, independent adult.

When your child leaves home to get a job, go to college, serve in the military, start a business, or begin raising a family, they'll need self-confidence, life skills, and the mental capacity to deal with what life will throw at them.

This book has been about that mental capacity—how a young person can deliberately build an impressive foundation of basic thinking skills. To take advantage of this opportunity, teens need to grasp the facts that have been the subject of this book:

- The prefrontal cortex (PFC)—the smart part of the brain—is the last area to finish connecting circuits for foundation skills.
- Teen PFC development begins immediately before puberty with a second blossoming of brain cell connections.
- The foundation construction process, which must happen during the adolescent years, proceeds the same way that other areas of the brain developed earlier in life: blossoming, pruning, and myelination.
- When teens exercise basic critical thinking skills, they stimulate the brain cells in the PFC to connect.
- To make the circuits more efficient, connected cells are insulated with myelin, and unused connections are pruned away.
- Only the child can do the work to make the connection.
- Quite a few academic courses, extra-curricular activities, and games work to wire the PFC.
- Some teens get lucky: they choose to get involved in the right activities by chance.
- Teens may not be motivated to do the work because (1) they aren't aware of what the PFC does or how it

develops, and (2) the process is slow, silent, and unseen, so there are no obvious signs of progress.

- The opportunity window for basic development closes when there are no more cells and connections to be pruned—near the end of adolescence.
- During adolescence the developing PFC needs to be protected from alcohol, drugs, and excessive screen exposure.
- The consequence for failing to protect the PFC and exercise thinking skills during youth: limited adult intellectual capacity.
- Adults can help by informing, encouraging, and supporting.

With this information, young people can make conscious choices to develop thinking skills that will enable them to understand what's going on in the world, solve problems, and use good judgment.

While it's true that adults can't do the work for teens, there are four ways for adults to be a positive influence:

1. Share what you've learned about your teen's brain. Make sure your adolescent child understands that basic development of the "smart" part of his or her brain is happening right now. Give them information about how this process works, that their adult brainpower will depend on how much work they do to exercise it, and that the time window for making a difference is limited. *The best way to do this is to encourage them to read this book.*

2. Help protect their PFC. Alcohol, drugs, addictive action video games, and excessive screen time will introduce chemicals into the brain that can disrupt the process. Make sure your child knows about these dangers, and establish boundaries that support healthy development.

3. Stimulate thinking. Instead of instructing, explaining and giving answers and solutions, ask the kind of open-ended questions that will encourage them to think for themselves.

4. Encourage and support learning and activities that exercise the PFC. Developing a superior mind doesn't require special exercises or more work added to your child's already full plate. No matter what, they'll go to class and get involved in activities. Many of the courses and activities that promote PFC development are interesting and fun. By understanding the benefits, teens can take courses seriously and choose activities knowing that they're also building brainpower.

Throughout, I've emphasized this point: *your child no longer has to get lucky*. Knowing what's going on and how it works, and with the motivation to do so, your child can take the bull by the horns and make it happen.

But will they accept your encouragement? Will they read the book? Will they care enough about growing a smarter brain to follow through on its suggestions?

I hope this book has convinced you that there's a lot you can do to make the best-case scenario happen. In fact, if your child hasn't reached puberty yet, they can begin to learn executive function and critical thinking skills during the grade school years. *Wiring these circuits before puberty can make things easier during adolescence*. Then, when the second wave of over-production of brain cells and connections happens in the PFC, the predictable fog of early adolescence is likely to be less disorienting.

Realistically, much of your success will depend on the relationship you've developed during the years leading up to adolescence. As a result of my ongoing study of teens and parenting, nurturing a strong bond with your child depends largely on what I consider the *fundamentals of effective parenting:*

1. Give unconditional love. Parental love is instinctive, but it can be eroded. Your child is going to make you angry. There will be the regular, petty irritants and disappointments that make you want to say or do things you'll regret. These reactions make a bad situation worse, doing damage to a child's already fragile self-esteem. In the worst case, someday your child might shock you by doing something unthinkable, such as hurting someone, becoming addicted, breaking the law, or getting pregnant. As you deal with things that make parenting hard, what your child will need most, in every case, is knowing that your profound love never wavers, is always there. This means expressing your love convincingly, even at the most troubling moments. As you help get your child back on track, your love will look like empathy, compassion, respect, understanding, tolerance, and forgiveness.

2. Be present. At home. At play. At games. At practices. At recitals and performances. When your child comes to you with a question or something to say, give your full attention as often as possible, for as long as possible, because opportunities to build your parent-child relationship will come and go, and time slips away. Be present emotionally as well as physically, with your attention focused on your child in the moment. You can't coach and teach if you're not there. You can't talk about the big issues if you haven't built a trusting relationship. The wounds caused by years of being distant or unavailable are hard to heal.

3. Keep the end in mind: you're preparing your child to be a happy, independent adult. The onset of puberty is the beginning of something new, and you'll need to change the way you see your child. The best experts affirm that you'll no longer be caring for a little kid. You'll be raising an adult, helping your adult-in-the-making prepare for life beyond the home.

The goal during adolescence is to help your child ingrain values, life skills, character strengths, wisdom and lessons that come from experience. This means young people don't need to be protected from problems, stress and failure; they need lots of opportunities to practice dealing with it.

4. Parent by example. Your child might believe some of what you say, but his or her constant go-to guide is observing what you do. You know that "do as I say, not as I do" doesn't work. Sometimes a parent's behavior is so hurtful that a child will say "I'll never be like that." But for many years your child has been in awe of you, and your example will nearly always trump everything else. Don't think your child will get excited about reading if you don't read. You can't expect your kid to keep a cool head if you often lose your temper. Your child is unlikely to be kind and generous if you're neglectful, manipulative, or abusive. As your teen learns to drive, don't expect him to come to a complete stop when you habitually roll through stop signs. If you abuse alcohol, do you really think your child will become an adult who drinks in moderation? If you're practical and careful with money, your child is likely to take a similar approach as an adult. It's a tall order, but you need to be the kind of person you want your child to become.

5. Improve your communication skills. A parent-child relationship—whatever it becomes—is built on the quality of communication. Everyday interactions can go badly or they can go well, depending on how well you communicate. The problem is, few parents have had training in effective communication skills. Most of us learned to deal with each other instinctively, "on the street," so to speak, and reacting to your child's behavior on this level can eventually push the two of you apart. Anger, put-downs, criticism, lecturing—these primal reactions alienate teens. There's a lot you can teach your child, but these talks may never happen if they feel that talking to you doesn't work.

The most important communication skill is *listening*, which is a component of several other important parent-child communication skills, such as giving and receiving *feedback*, including praise and constructive feedback; *dialogue*; and *conflict resolution*. In a way, effective communication is like the game of chess, because you can work on improving your game all your life, and you'll never come to the end of mastery. But you can get very good at it. The journey towards more effective communication is worth the effort; it's how lasting relationships are formed.

6. Help your child build a strong work ethic. When I was 15 I mowed my aunt's lawn with a borrowed push mower in the middle of a Kansas summer. Whew! She was so impressed with my effort that she gave me $5, which is the equivalent of over

$40 in today's money. As a teen, I caddied to earn money for what I wanted.

When a close friend's father died when he was 15, he had to take over managing the family farm, in addition to class and sports. The bad news: it was tough for him. The good news: he established an amazing work ethic. Today he's a senior executive in the oil and gas industry, and he can fix literally anything, even without the right tools or spare parts. I've seen him do it many times.

In the world of work, organizations want young adults who take initiative, make an effort, care about excellence, and stick with a task until it's done. They want entry-level employees who dig in and become part of the solution. They don't want self-centered, laid-back, complaining kids who think they're entitled to a promotion for showing up. Young people don't inherit a strong work ethic, and it doesn't suddenly awaken in them when they leave home. It comes from establishing strong work habits while they're young. You can help by arranging opportunities to work and encouraging your child to persist through the tough spots.

7. Nurture strong self-esteem. Many of the unfortunate things that teenagers do happen because they're vulnerable to peer pressure. They aren't sure they're worthy of being noticed, liked and accepted, so they sometimes go along with behavior that can get them in trouble. They're nervous about saying no because they aren't sure who they are or what they stand for. Doing dumb things is a part of growing up; it's normal and inevitable. But parents sometimes react by berating their kids for making mistakes, instead of standing with them and helping them *learn from mistakes*. It's hard to remember that young people have so much to learn and such a long way to go before they discover wisdom. For the sake of their future happiness, they need parents' help to survive the pitfalls of adolescence while building confidence and a strong sense of self.

8. Support your child's passionate interests. I believe it's important for kids to be excited about something, but hopefully not something hurtful or addictive. Being intensely committed to something can ultimately be the driving force toward a rewarding life. You might think your kid is too old for Legos, or that the dance class is nice but not going anywhere. You may wish your child was focused on photography, chess or computer programming instead. Just show interest and add your enthusiasm to that of your child. It's often hard to know how important an activity can be developmentally, and passionate interests change. If your kid catches fire about a particular activity, show your support and count yourself lucky.

9. Share what you've learned. As I write this, many of today's young adults are attending "adulting" classes to learn basic skills their parents could have taught them, but didn't. Stuff like money management, cooking, and home maintenance. I once knew a young woman who didn't know anything about cooking. She couldn't even fry an egg, *although her mother was a gourmet chef!*

I grew up in a large family, and I think my parents were simply overwhelmed by the daily challenge of feeding, clothing and protecting eight kids. They passed on very few practical skills or life wisdom, and when I found myself out in the world I quickly realized I had a lot of catching up to do.

Kids need to learn so much to prepare for life, and most of this isn't taught in schools. They need to acquire values, character strengths, practical life skills, relationship skills, critical thinking skills, and wisdom. That's a lot, and in the best case you don't leave this to chance. Don't do for your kids what they can do for themselves. Teens can do their own laundry, cook meals and do other family chores. If you fix your own car or do gardening, you can ask your teen to help. Teach them what you know. A child who becomes a self-reliant adult is a fortunate individual.

10. Encourage your child to think. This is what this book has been about. Teenagers may sometimes do silly things; most parents believe this is a youthful phase they'll outgrow by the time they're adults. But good judgment, critical thinking, and executive function are *skills*. Like all skills, they have to be acquired through practice.

In other words, young people learn and use concepts, analyze, think logically and solve problems by doing these things, over and over. The kids who are coached to do a lot of this kind of thinking will grow up with more advanced thinking skills. Those who aren't, may not. Kids can get lucky if teachers prod them to use these skills or if mentors ask the kind of questions that stimulate thought. On the other hand, parents who give the answers and solve problems for their kids rob them of opportunities to develop the ability to think for themselves. The foundation for PFC development has to be laid down during adolescence.

If this doesn't happen, a young adult may have to make do with some serious limitations. Look around you. You know good, hardworking people who can't tackle difficult problems and careers. I wrote Chapter 5 to help parents ask the kind of questions that will cause their child to do their own thinking.

As a concerned, conscious parent, you're on a dedicated learning journey while managing the miracle of raising an independent adult. When your child becomes a teenager, hopefully they'll welcome your input and encouragement.

If your child is already a teenager, it's not too late to make adjustments. You may have some explaining to do, but if you share the facts about their developing brain, and if you work together to put these commonsense recommendations into practice, some wonderful things can happen.

Major takeaways from Chapter 11:

- No one is a perfect parent, but you can learn from each experience.
- Preparing for adolescence can make parenting a teen a lot easier.
- The best way to prepare your child for adolescence is to begin now to implement the recommendations of this book, and stay true to the fundamentals of effective parenting.
- Working on improving communication skills is at the top of the list.

Recommended actions for parents:

- From time to time, review the list of ten fundamentals to focus on a learning opportunity.

- While this book may fall in the parenting category, you should definitely encourage your child to read it. It's brief, written so teens will understand it, and the insights can open doors to future success.
- If your child is already a teenager, the insights of this book may inspire you to want to undo and redo some things. Don't hesitate to have give-and-take conversations about these issues with your teen.

BONUS: Go to https://DrDennyCoates.com/bonus to download a free guide: "The No. 1 Way to Nurture the Bond with Your Teen: Become an Awesome Listener."

AFTERWORD

The "big idea" of this book is that without any special exercises, young people can *consciously* wire their prefrontal cortex (PFC) for a host of powerful thinking skills. The benefits of doing this are enhanced judgment, problem solving and decision-making skills, and self-regulation—all of which can help keep a child safe from the gauntlet of adolescent problems and tragedies, and then later empower a young adult to greater success in any profession.

This is an astonishing new opportunity. Because it's based on research reported only 20 years ago, it's one that wouldn't have been possible for previous generations of adolescent children.

More important, the ability to *deliberately* build cognitive capacity is available to *any child*, regardless of gender, culture, race, or economic background.

All that's required is the willingness to learn about the importance of the PFC, how basic brain development happens, and which learning and extracurricular activities best develop the smart part of the brain. And the first step is easy: all a kid has to do is read this book, discuss it with an adult, and decide if he or she cares enough to follow through.

This means the opportunity isn't limited to the usual cohort of honor roll students and high-achievers. It's equally available to underprivileged or at-risk children. It's now possible for them to do what only a few fortunate pre-21st century kids happened to do: develop the kind of mind that leads to success in practically anything.

This is an unusual book—one promoted to adults but written to be accessible to youth, a book that can realistically be a difference-maker in millions of lives. If you share my desire to bring this opportunity to as many kids as possible, then I ask you to share what you've learned. If you know parents of preteens or teens, tell them about the book. If you're a teen, you may know a friend who could benefit from reading the book.

Also, I waive my speaking fee for not-for-profit groups. Interested adults can get information about the book on Amazon.com, https://SmarterTeenBrain.com, or my website, https://DrDennyCoates.com. If you know of an influencer I should contact, please reach out to me at drdennycoates@gmail.com.

Appendix A

A Selective, Annotated Bibliography of Other Books about the Adolescent Brain

If the book in your hands has sparked an interest in learning more about the adolescent brain, about two dozen other books on this topic have been published since the turn of the century. Some of them are especially helpful resources for parents of teens.

When looking for information about science, it helps to know there are three kinds of books.

The first are summaries of research reports, usually written by brain scientists or university professors. These reports and books are formal, highly technical, and of interest mostly to other scientists—not the general public. An example is *The Adolescent Brain: Learning, Reasoning, and Decision Making,* edited by Valerie Reyna and others.

The second level is science writing, produced by knowledgeable people who do a thorough "review of the literature" and attempt to make a science topic understandable to people like you and me. The books listed below fall into this category.

The third level includes pop science journalism and "pseudoscience." These media pieces, videos, and sometimes books are created by people who may not have immersed themselves in the science, but who sample science writing and other sources to introduce the public to newsworthy topics. Because getting the content produced is more urgent than mastering the subject matter, they often misunderstand the science or draw erroneous conclusions.

Of course there are hundreds of books about parenting and dozens of books about raising teens. Many of these are wonderful resources, such as Jane Healy's *How to Have Intelligent and Creative Conversations with Your Kids*; but my focus in this list is the adolescent brain, not the broader challenge of parenting teenagers.

My book is based on three decades of study, including the publications from the first two levels. The books below introduce current thinking about the developing adolescent brain. All of them are based on a reliable understanding of the science. However, only one of them, *The Power of the Adolescent Brain*, written for teachers, contains recommendations for developing the adolescent brain.

To help you decide where to start, I've included my descriptive comments.

Walsh, David. *Why Do They Act That Way: A Survival Guide to the Adolescent Brain for You and Your Teen,* 2nd ed. Atria, 2014.

One of the strengths of this book is how consistent it is with teen brain research. Also, it's written from the perspective of a family psychologist, so it touches on the most important issues and offers experience-based insights.

Armstrong, Thomas. *The Power of the Adolescent Brain: Strategies for Teaching Middle and High School Students*. ASCD, 2016.

This book contains an excellent science-based summary of adolescent brain development. It's especially valuable because it's the only other book I'm aware of that recommends strategies for developing the adolescent prefrontal cortex.

Clarvier, Ron. *Teen Brain, Teen Mind: What Parents Need to Know to Survive the Adolescent Years*, 2nd ed. Key Porter, 2009.

Clavier is another author who has done his homework about adolescent brain development. He uses an entertaining style to suggest helpful insights and strategies for raising teenagers.

Jensen, Frances. *The Teenage Brain: A Neuroscientist's Survival Guide to Raising Adolescents and Young Adults*. Thorson's, 2015.

Written by a brain scientist, this book is an authoritative guide to the whole brain, not just the developing prefrontal cortex. It features lots of references to research, quite a few stories about parents and teens, and some useful recommendations.

Dunckley, Victoria. *Reset Your Child's Brain: A Four-Week Plan to End Meltdowns, Raise Grades, and Boost Social Skills by Reversing the Effects of Electronic Screen Time*. New World, 2015.

For over a decade, Dunkley, an integrative psychiatrist, has been resolving severe behavioral issues caused by excessive screen exposure. Her recommendations are based on a thorough understanding of the science of teen brain development. Readers seeking more information about the dangers of excessive screen time can consult her extensive chapter endnotes.

Siegel, Daniel. *Brainstorm: The Power and Purpose of the Teenage Brain*. Tarcher, 2013.

In addition to insights about the teen brain, his book features recommendations for dealing with a wide range of teen issues. Siegel's strategies mirror his professional focus: that of a psychiatrist who specializes in mindfulness.

White, Aaron and Scott Schwartzwelder. *What Are They Thinking: The Straight Facts about the Risk-Taking, Social Networking, Still-Developing Teen Brain*. Norton, 2013

A biological psychologist, White touches on lots of teen issues, focusing mostly on advice on how to protect the brain from the effects of brain-damaging substances and screens.

Appendix B

Additional Reading: Books about Parenting Teens

There's a lot more to raising young adults than helping them grow a smarter brain.

In addition to the books cited in Appendix A, which focus on the developing adolescent brain, I share here a list of books about the broader subject of parenting teens (which is a subset of the even broader genre of books about parenting). These books address the many challenges that parents face as they try to prepare their kids for adulthood: sex, pornography, dating, relationships, self-identity, peer pressure, risk-taking, bullying, depression, self-harm, suicide, financial literacy, community service, family chores, self-management, safe driving, life skills, high school graduation decisions, and more.

Instead of a comprehensive bibliography, here is a short list of books I found both enjoyable and helpful:

Barkley, Russell, and Arthur Robin. *Your Defiant Teen: 10 Steps to Resolve Conflict and Rebuild Your Relationship.* New York: Guilford Press, 2008.

Efessiou, Christos. *CDO – Chief Daddy Officer: The Business of Fatherhood*. Charleston, SC: Advantage Media, 2011.

Faber, Adele, and Elaine Mazlish. *How to Talk So Teens Will Listen & Listen So Teens Will Talk*. New York: HarperCollins, 2005.

Gilboa, Deborah. *Get the Behavior You Want...Without Being the Parent You Hate: Dr. G's Guide to Effective Parenting*. New York: demosHealth, 2014.

Ginott, Haim G. *Between Parent and Teenager*. New York: Avon Books, 1971.

Guare, Richard, et al. *Smart But Scattered Teens*. New York: Guilford Press, 2013.

Gurian, Michael. *The Wonder of Girls: Understanding the Hidden Nature of Our Daughters*. New York: Pocket Books, 2002.

__________*The Wonder of Boys: What Parents, Mentors and Educators Can Do to Shape Boys into Exceptional Men*. New York: Penguin Group, 1996.

Healy, Jane M. *Failure to Connect: How Computers Affect Our Children's Minds—and What We Can Do About It*. New York: Touchstone, 1998.

Lythcott-Haims, Julie. *How to Raise an Adult: Break Free of the Overparenting Trap and Prepare Your Kid for Success*. New York: Henry Holt, 2015.

Kersting, Thomas. *Disconnected: How to Reconnect Our Digitally Distracted Kids*. 2016.

Kolari, Jennifer. *You're Ruining My Life! (But Not Really): Surviving the Teenage Years with Connected Parenting.* New York: Penguin Group, 2011.

Riera, Michael. *Uncommon Sense for Parents with Teenagers.* Berkeley, CA: Ten Speed Press, 2012.

__________*Staying Connected to Your Teenager: How to Keep Them Talking to You and How to Hear What They're Really Saying*. Boston: DaCapo Press, 2003.

Rosemond, John. *Teen-Proofing: A Revolutionary Approach to Fostering Responsible Decision Making in Your Teenager*. Kansas City, MO: Andrews McMeel, 1998.

Sells, Scott P. *Parenting Your Out-of-Control Teenager: 7 Steps to Reestablish Authority and Reclaim Love*. New York: St. Martin's Griffin, 2001.

Simon, Michael Y. *The Approximate Parent: Discovering the Strategies That Work with Your Teenager*. Oakland, CA: Fine Optics, 2012.

Trittin, Dennis, and Arlyn Lawrence. *Parenting for the Launch: Raising Teens to Succeed in the Real World*. LifeSmart, 2013.

[illegible] (But Not Really) [illegible] Connected Parenting. New [illegible]

[illegible] for Parents with Teenagers. [illegible] Press, 2011.

———. [illegible] Your Teenager: How to Keep Them [illegible] to Hear What They're Really [illegible] Press, 2005.

Rosemond, John. [illegible]: A [illegible] Approach to [illegible] Making [illegible] Your Teenager [illegible] Andrews McMeel, 2001.

[illegible] Your [illegible] Teenager [illegible] New York [illegible]

[illegible] Discovering the [illegible]

[illegible]

Dr. Denny Coates is a parenting thought leader who writes about adolescent brain development for parents and adults who work with youth. His focus is to provide science-based insights to help adults guide young people to develop their prefrontal cortex for executive/critical thinking skills in order to avoid the pitfalls of growing up and achieve optimum intellectual capacity for adult life. As CEO of Performance Support Systems, Inc., Dr. Coates has created award-winning brain-based programs and resources for parents, leaders and teams for over 30 years.

A graduate of West Point (1967), he retired from the U.S. Army as a lieutenant colonel (1987). He earned his Ph.D. from Duke University (1977) and served on the faculties of the United States Military Academy, the Armed Forces Staff College, the College of William and Mary, Thomas Nelson Community College, and the Center for Creative Leadership.

Dr. Coates is the creator of several online coaching and reinforcement programs for building communication skills and personal strengths:

- Strong for Parenting – for parents
- Strong for Families – for family therapists, school counselors and other helping professionals
- Strong for Life – for teenagers
- Strong for Performance – for adults in the workplace
- 20/20 Insight Gold – online individual, team and organization performance feedback surveys

He is also the author of:

- *Conversations with the Wise Aunt*—for middle school girls

- *Conversations with the Wise Uncle*—for middle school boys
- *The Sacred Purpose*—for administrators of youth sports programs
- *The Dark Secret of HRD*—for business executives

Website & blog: https://DrDennyCoates.com

LinkedIn: Dr. Denny Coates

Twitter: @DrDennyCoates

Facebook author fan page:
https://www.facebook.com/DrDennyCoates